NATIVE PLANTS
of *Melbourne*
and Adjoining Areas

A Field Guide

David & Barbara Jones

Bloomings Books

Published by Bloomings Books

BLOOMINGS BOOKS

37 Burwood Rd
Hawthorn Victoria 3122
Australia
Phone: +61 3 9819 6363

Bloomings Books is a specialist publisher and wholesaler of natural
history and horticultural books.

The National Library of Australia Cataloguing-in-Publication Data:

Jones, David L. (David Lloyd), 1944-

A field guide to the native plants of Melbourne.

ISBN 1 876 47313 4

1. Botany - Victoria - Melbourne. I. Jones, Barbara, 1943-

II. Title

581.99451

Design: Andrew Rankine Design Associates

Editor: Anne Findlay

CONTENTS

DEDICATION

To the memory of Dr James Hamlyn Willis (1910–1995), botanist extraordinaire, inspiring orator, author, gentleman and friend for over 30 years.

PREFACE

The idea for this book originated during a phone conversation with Warwick Forge of Bloomings Books. The suggestion was for a simple field guide to the indigenous plants of the Melbourne region—we took it as a challenge. Having spent some 34 years of our lives in Melbourne, and for much of that time seeking out native plants, we felt qualified for the task and we hope we have succeeded.

The flora of the greater Melbourne region contains some 1150 species. It was not possible to include all of them in this publication so a suitable compromise was to cover a selection of the prominent or distinctive indigenous species. We have included some species which are now rare or uncommon through no fault of their own. We feel their inclusion adds interest and emphasises the current plight of many species, not only in this region, but also in other parts of the state or the nation.

Information on native plants has burgeoned in the last two decades, with much of the knowledge being compiled by members of organisations such as the Australian Society for Growing Australian Plants (ASGAP). More recently the new *Flora of Victoria* provides a detailed scientific compilation of the state's flora.

One monumental publication deserves special mention—the *Flora of Melbourne* (Society for Growing Australian Plants Maroondah, Inc.) published in 1991, at the time of writing unfortunately out of print. No praise is too high for the band of dedicated and enthusiastic people who compiled this significant production over about seven years. In many ways the very compilation of such a fine work by enthusiasts reflects the changed attitudes in the community to natural issues and the environment.

In preparing this book we have made every effort to keep up with name changes which follow botanical research, not always an easy task. Some of the names used in this book may have changed by the time it is published.

An alphabetical layout has been used throughout and there has been no attempt at taxonomic grouping. The use of botanical terms has been kept to a minimum and a glossary is provided to explain the few that have been used. Common names are included for all species. The book is illustrated by 252 colour photos, the majority of which we took ourselves. Contributions are credited in the acknowledgments.

INTRODUCTION

The year 1835 was a pivotal one for the flora of the Melbourne region, for in May–June of that year John Batman negotiated the purchase of a huge parcel of land from the local Aboriginal tribe, and in August of the same year, a party initiated by John Fawkner landed on the northern bank of the Yarra River. A steady influx of land-hungry squatters followed these pioneers and, while this resulted in the construction of the great city of Melbourne, it also brought about fragmentation of the local flora and extensive destruction of the plant communities in some areas. So complete is this destruction in many suburbs, that nothing remains to indicate what has been lost. In other suburbs precious relicts of the original flora still exist. Today many of these relict areas are protected as reserves, but still the threats and changes continue as small pieces are hived off for development, and depradations of pests (such as introduced slugs and snails) and vigorous weeds invade to forever alter these priceless vestiges.

Boundaries of the Melbourne region

As defined by the Twenty-Seventh Conference of Statisticians of Australia in 1965, the urban area of Melbourne extends some 20 kilometres to the west and north and 25 kilometres to the east and south-east. However, Melbourne's urbanisation has extended its tentacles much wider than these limits and for the purposes of this book it seems more logical to extend the boundaries to include the outer suburbs of major settlement (see map on front cover). To the south-west this encompasses the Bellarine Peninsula, to the north, Wallan and Whittlesea, to the north-east Kinglake and Healesville, the east includes the Dandenong Ranges, and the south-east all of the Mornington Peninsula extending to the eastern shores of Westernport Bay.

Melbourne's disappearing plant communities

The Melbourne flora has been changed so radically in the 164 years since white settlement that it is often difficult to envisage the natural communities of plants which covered the various districts. In some regions the modification has been so drastic that historical records and scant relicts provide the only clues to the original plant cover. The large flat to undulating basalt plain to the immediate west and north-west of Melbourne, with its thin topsoil and heavy clay subsoils, provides a good example. The original vegetation of this area must have been a boon to the early squatters since it was dominated by grasses, and the trees and shrubs were largely confined to rocky outcrops, drainage lines and streams. With limited clearing needed and few impediments to grazing it is no wonder that so little of it remains today.

Urbanisation, resulting from people's strong desire to live near the coast, has been the major cause of the fragmentation and destruction of the vegetation fringing the eastern shores of Port Phillip Bay and adjacent inland areas. We can only wonder at the losses when reading historical articles on the springtime floral displays in the extensive heathlands that flourished around Sandringham, Black Rock and Beaumaris. Some of these narratives include tantalising records of orchids, the identity of which can only be guessed today. Even in the last 30 years we have witnessed the destruction of the fabulous wildflower gardens that proliferated on the red sandy areas to the east of Frankston. We can also speculate on what interesting aquatic plants inhabited the apparently extensive swamp systems that once existed around Carrum and Edithvale.

In other areas significant samples of the vegetation remain in relict patches and reserves. This is certainly the case in some of the outer eastern suburbs and the Dandenong Ranges, although certainly the changes in these areas have been massive. Of strategic significance has been the loss of 'hot spots', those priceless parcels of land, often small, which supported tremendous local diversity, including an abundance of showy flowering shrubs, forbs, and large numbers of orchids. Were's paddock, Greensborough, was one such site with which we were familiar, as also was the land adjacent to the railway station at Boronia, the railway reserve between Heathmont and Bayswater, the land around the Eltham football ground, and numerous patches of bush in Reynolds Rd, Donvale and Tindals Rd, Park Orchards.

Orchids are a special love of ours and on numerous occasions we have noted the close relationship that exists between an abundance of these special plants and a general health and diversity of the flora at a particular site. Thus in our lifetime we have seen the disappearance of numerous orchid-rich patches of bush in such suburbs as Bayswater, Beaumaris, Berwick, Boronia, Braeside, Croydon, Deer Park, Diamond Creek, Donvale, Eltham, Ferntree Gully, Frankston, Greensborough, Heathmont, Langwarrin, Lysterfield, Mitcham, Ringwood, Rye, Scoresby, Seaford, Templestowe, The Basin, Warrandyte and others. Even Beckett Park, in the well-established Melbourne suburb of Balwyn, contained five species of orchids in the early 1950s, as well as a host of other herbs and forbs.

ABOUT THIS BOOK

Choice of species

Various criteria have been considered when choosing which plant species to include in this book. Those with prominent features, such as conspicuous flowers, or distinctive plants which are familiar or readily recognisable were obvious choices. Similarly, species widely distributed in the region were also considered strong candidates, and we tried to include representatives from as many genera as possible. By the same reasoning, an effort was made to include representative species from suburbs to the west, north and east of Melbourne, and the remaining coastal fringes. If there is a bias it probably leans towards the east, simply because so little habitat remains on the western side. We also tried to cover major plant groups such as ferns, grasses, sedges, lilies and orchids, as well as shrubs and trees. A few rare species, or those with narrow distributions within the region, have been included because we feel they have an interesting story to tell. With less than a quarter of the Melbourne flora covered there will obviously be significant omissions, but we hope we have provided a suitable representative coverage of the subject.

Plant names

Currently accepted plant names have been used throughout this work and any recently changed names or well-known older names are included for convenience as synonyms in the notes and in the index. Common names are included for all species and the plant family is given to indicate relationships between various genera. Occasionally a subspecies or varietal name is used to indicate that a distinctive variant of the species occurs in the region.

Photographs

All species are illustrated by photographs and in some cases an accessory small photograph depicting a feature useful for identification is included as an insert on the main subject. Some of the photographs were not taken in the Melbourne region, but are included because they are representative of the species.

Description, similar species and flowering period

A simple description of the plant is included using non-scientific terms where possible. This is biased towards features of the plant which, in combination with the photograph, will aid in its identification. Some measurements are included but these are always at the upper limits and it must be realised that tremendous variation is common in growth features such as plant height and leaf and flower dimensions.

Other species in the region which are likely to be confused with the plant in question are distinguished under the heading 'Similar species'. These plants are often not illustrated but it is hoped that the inclusion of simple distinguishing features will help identify the subject in hand.

Flowering times can be a useful aid to identification as some plants, such as the orchids, have a restricted flowering period. Many other plants, however, have extended flowering periods and groups such as lilies and grasses can respond sporadically to unseasonal rains with subsequent growth and flowering well out of their normal time.

Distribution

The occurrence of the plant within the greater Melbourne region is given as a general suburban distribution rather than specific localities. Many of the records on which distributions are based are very old and the species may now be extinct in many of these sites. Following the regional distribution is a generalised note on the occurrence of the species within the state of Victoria and its distribution elsewhere in Australia and sometimes overseas.

Habitat

We have deliberately used general terms to provide habitat details because in our experience people relate to these more readily than they do to the very specific terms used by ecologists. Definitions of the habitats mentioned are included in the glossary. It is not the aim of this book to cover the plant communities of Melbourne in any detail. Any reader who is interested in this latter subject should read the fascinating account by Roslyn Savio entitled 'Melbourne's Indigenous Plant Communities' which is included in *Flora of Melbourne*, 1991 (Society for Growing Australian Plants Maroondah, Inc).

Cultivation

We hope that having assisted people to identify local plants via this book, we can encourage them to grow species indigenous to their local area in their own gardens. Notes have been included indicating whether a species is readily adaptable to cultivation and briefly mentioning basic requirements. These comments are deliberately limited and further information on the subject can be obtained from a host of specialist native plant books. It should also be mentioned that nowadays some Melbourne nurseries stock local indigenous plants.

Abbreviations

ACT	Australian Capital Territory
NSW	New South Wales
NT	Northern Territory
NZ	New Zealand
NCal	New Caledonia
PNG	Papua New Guinea
Qld	Queensland
SA	South Australia
Tas	Tasmania
WA	Western Australia
aff.	affinity with
alt.	altitude
cm	centimetre
m	metre
mm	millimetre
ssp.	subspecies
var.	variety

ACKNOWLEDGEMENTS

We are especially grateful to Anne Mentiplay, Helen and John Donnell, Rodger and Gwen Elliot and Sandie and Rob Ogden for hospitality at various times. We thank Tim and Emma Jones for minding the home patch in our absence.

We also thank the following people for contributing photographs to the book: Mark Clements (pp. 45, 51, 54, 63, 75, 87, 92, 117, 151, 186, 200, 202, 222, 236); Rodger Elliot (pp. 9, 38, 69, 74, 136, 143, 160, 168, 174, 188, 193, 194, 212, 213, 218, 228, 232, 234, 246, 250); John Fanning (pp. 118, 208, 209); Everett Foster (pp. 49, 76); Jeffrey Jeanes (pp. 47, 48, 50, 64, 80, 89, 101, 119, 122, 170, 177, 205, 201, 206, 236, 238); Jean Johnson (p. 77); and Les Rubenach (pp. 12, 210).

David and Barbara Jones
Canberra, April 1999

Family Mimosaceae

Description

Spreading shrub, to 2.5 x 3 m, with long arching branches and small, oblong to ovate, green phyllodes, to 2 cm x 1 cm, ending in a point that curves back on itself. Bright golden yellow, globular flower-heads, arising from the axils, produce a massed floral display, and are followed by straight or curved pods, to 5 cm x 0.3 cm.

Flowering period	Aug. to Nov.
Distribution	Northern and eastern suburbs, and throughout much of the state; also NSW and SA.
Habitat	Open forests, particularly those dominated by various box eucalypt species, especially in stony or shaley soils.
Notes	A very showy wattle with fragrant flowers.
Similar species	None in the region.
Cultivation	Easily grown in a sunny position in well-drained soil.

Family Mimosaceae

Description

Bushy tree, to 30 m tall, with silvery bark on the young branches, and silvery hairy new growth. The bipinnate leaves, to about 12 cm long, have 10–20 major divisions, each bearing numerous silvery-grey to bluish leaflets, to 0.5 cm x 0.1 cm. Pale yellow to bright yellow, globular flower-heads, borne in large terminal racemes, are followed by silvery to bluish pods, to 10 cm x 1 cm.

Flowering period	Aug. to Oct.
Distribution	Widespread in the Melbourne region and throughout much of the state; also NSW, ACT and Tas.
Habitat	Moist forests, woodland and streamside vegetation.
Notes	A fast-growing species sometimes forming thickets by sucker production. Very showy when in flower and the flowers are attractively scented. The seeds and gum were eaten by the Aborigines.
Similar species	*A. mearnsii* has dark green leaflets.
Cultivation	Very easily grown in a moist position but must have room to develop.

Family Mimosaceae

Description

Erect to spreading shrub, to 3 m x 3 m, which often has an open habit, with narrow 4-sided dark green prickly phyllodes, to 3 cm x 0.3 cm. Cream to slightly yellowish globular flower-heads are produced from the axils and are followed by straight or curved pods, to 12 cm x 0.5 cm.

Flowering period	Mainly Jan. to May, but often also with a spring flush.
Distribution	Eastern suburbs, Dandenong Ranges, and common throughout much of the state; also NSW, ACT and Tas.
Habitat	Open forest, often on stony ridges but also on heavier soils of the lower slopes and valleys.
Notes	Formerly well known as *A. diffusa*.
Similar species	None in the region.
Cultivation	Easily grown in a range of soils and positions.

Family Mimosaceae

Description

Fast growing shrub or small tree, to 15 m x 5 m, with an open to bushy habit and large, sickle-shaped, bright green phyllodes, to 20 cm x 2 cm. Cream to pale yellow, globular flower-heads are carried in axillary racemes, and are followed by clusters of strongly curved, light brown pods, to 20 cm x 0.7 cm.

Flowering period	Dec. to March.
Distribution	Widespread in the Melbourne region and throughout much of the state; also Qld, NSW and ACT.
Habitat	Open forest, woodland, heathy forest and grassland.
Notes	A long-lived species valuable for revegetation purposes.
Similar species	*A. melanoxylon* grows much larger and flowers earlier (Sept. to Dec.).
Cultivation	Easily grown in well-drained soil in a range of positions, from shade to sun.

Family Mimosaceae

Description

Usually a dense shrub, to 3 m x 2 m, often with attractive reddish new growth. The broad, dark green phyllodes, to 8 cm x 3 cm, which taper to each end, have a prominent midrib and the margins are often yellowish. Cream to yellow, globular flower-heads, carried in racemes in the upper axils, are followed by curved, brown pods, to 10 cm x 1 cm.

Flowering period	July to Oct.
Distribution	Eastern suburbs, Dandenong Ranges, and throughout much of the state; also Qld, NSW, Tas, SA and WA.
Habitat	Open forest, woodland and heathy forest.
Notes	A variable species which often dominates the vegetation of an area, especially in the first five years after a fire.
Similar species	None in the region.
Cultivation	Easily grown in a range of soil types and positions.

Family Mimosaceae

Description

Variable from a shrub to a small tree, to 10 m x 5 m, with an erect
to spreading habit and sharply pointed, rigid, dark green phyllodes,
to 4 cm x 0.5 cm. Conspicuous, rod-like, bright yellow flower-
heads, to 3 cm long, are followed by nearly cylindrical, leathery
pods, to 8 cm x 0.5 cm.

Flowering period	June to Nov.
Distribution	Eastern and south-eastern bayside suburbs, and widespread in southern parts of the state; also NSW and SA.
Habitat	Coastal scrubs, heathy forests and heathland.
Notes	A very prickly species which is an important refuge plant for small birds.
Similar species	*A. verticillata* has much smaller, narrower phyllodes.
Cultivation	Easily grown in a range of soil types and positions.

Family Mimosaceae

Description

Bushy shrub or small tree, to 10 m x 5 m, with silvery bark on the young branches, and large, leathery green, often curved phyllodes, to 20 cm x 5 cm, each with a prominent midrib. Bright yellow to golden, globular flower-heads are carried in profusion on racemes in the upper axils, and are followed by straight, leathery, brown pods, to 13 cm x 2 cm.

Flowering period   Aug. to Oct.

Distribution   Widespread in the Melbourne region and throughout much of the state; also NSW, ACT and SA.

Habitat   Open forest, woodland and grassland, sometimes in dense stands.

Notes   Australia's floral emblem. Sap released from the trunk is an important food source for glider possums and was also eaten by the Aborigines.

Similar species   None in the region.

Cultivation   Easily grown in a range of soil types.

7

Family Mimosaceae

Description

Tall bushy shrub or tree, to 8 m x 5 m, with spreading or pendulous branchlets and narrow, bluish-green phyllodes, to 20 cm x 1 cm. Smallish, lemon-yellow flower-heads, carried in short axillary racemes, are followed by thin, brown pods, to 20 cm x 1 cm.

Flowering period	Mainly Nov. to March, but also sporadic at other times.
Distribution	Of scattered occurrence in many suburbs and often naturalising from cultivation; also Tas and SA.
Habitat	Coastal scrubs, swamp margins and streamside vegetation.
Notes	This species can sucker to form small clonal colonies.
Similar species	None in the region.
Cultivation	Popular for its fast growth and long flowering period. Adaptable and easily grown.

Family Mimosaceae

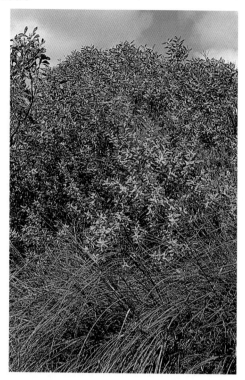

Description

Shrub or tree, to 8 m tall, with a mounded to widely spreading habit, and oblong to elliptical, dark green phyllodes, to 15 cm x 3 cm, with two to five prominent veins. Yellow, rod-like flower-heads, 2–3 cm long, borne in abundance in the axils, are followed by curved, twisted, brown pods, to 15 cm x 0.8 cm.

Flowering period	July to Oct.
Distribution	Eastern and south-eastern bayside suburbs and in southern areas of the state; also Qld, NSW, Tas and SA.
Habitat	Coastal scrub on dunes and headlands.
Notes	An important species for stabilisation in coastal districts. May colonise suitable areas if planted in habitats away from the coast. The green seeds were eaten by the Aborigines after cooking.
Similar species	None in the region.
Cultivation	Excellent in coastal districts, but can also be grown successfully in areas well away from the coast.

Family Mimosaceae

Description

Open shrub, to 3 m x 4 m, with erect to spreading, often willowy or weeping branches and narrow, greyish to bluish-green phyllodes, to 15 cm x 1 cm. Cream, sweetly fragrant, globular flower-heads, borne in axillary racemes, are followed by bluish, leathery pods, to 4 cm x 2 cm.

Flowering period	April to Oct.
Distribution	Eastern and south-eastern bayside suburbs and in southern parts of the state; also Qld, NSW, Tas and SA.
Habitat	Coastal scrub, heathland and heathy forest.
Notes	This ornamental species can have a long flowering period.
Similar species	None in the region.
Cultivation	Easily grown in an open sunny position in well-drained soil. Excellent for coastal districts.

Family Rosaceae

Description

Low, carpeting, prostrate groundcover with widely spreading, much-branched stems densely clothed with dark green, pinnate leaves, to 4 cm long. Each leaf has up to 11 toothed leaflets, to 1 cm x 0.4 cm. Erect, slender, leafless stalks carry globular, whitish flower-heads, which are followed by brown, spiny, globular, burr-like fruiting heads, about 2.5 cm across.

Flowering period	Oct. to Jan.
Distribution	Widespread in the Melbourne region and throughout much of the state; also Qld, NSW, ACT, Tas, SA and NZ.
Habitat	Grassland, damp to moist forests, woodland, and swamp margins.
Notes	The seed heads, which stick readily to clothing, shoelaces and animal fur, can be a nuisance but also provide a ready means of identification. Previously well known as *A. anserinifolia*.
Similar species	None in the region.
Cultivation	Easily grown in moisture-retentive soil.

Family Orchidaceae

Description

Small terrestrial orchid with a heart-shaped leaf, to 3 cm x 2 cm, which is dark green above, purplish-red beneath, and held above ground level on a slender stalk. Dark purplish-red flowers, with filamentous segments, to 4 cm long, are carried in slender racemes of up to nine flowers. The lip is smooth with a concave base and pointed apex.

Flowering period	July to Oct.
Distribution	Northern and eastern suburbs, particularly in the foothills and ranges, and eastern and western parts of the state; also Qld, NSW, Tas and SA.
Habitat	Open forest and heathy forest.
Notes	This species grows in loose, open colonies. The leaves are often more noticeable than the flowers, which exude a strong musty smell on warm days.
Similar species	None in the region.
Cultivation	Grown in containers by orchid specialists, but can be difficult to maintain.

Family Adiantaceae

Description

Ground fern with wiry, spreading, underground rhizomes forming clumps, sometimes many metres across. Tufts of finely divided, lacy, bright green fronds, to 50 cm x 20 cm, can be erect, arching or drooping. Each frond has a black stem, is divided two or three times, and has numerous, membranous, wedge-shaped segments with the margins shallowly lobed.

Distribution	Widespread in the Melbourne region and throughout much of the state; also Qld, NSW, ACT, Tas, SA, WA and some overseas countries.
Habitat	Moist sheltered sites in a range of habitats; also in moist to wet seepage areas in exposed sites and on embankments.
Notes	Plants grow rapidly during autumn and spring.
Similar species	None in the region.
Cultivation	Easily grown in containers or a partially sheltered garden situation in moist well-drained soil.

Family Lamiaceae

Description

Perennial herb forming clumps or colonies and spreading by suckers. Plants have a basal rosette of greyish-green leaves, to 12 cm long, with conspicuously wavy, toothed or lobed margins. Erect fleshy stems, which have similar, but smaller leaves, end in leafy spikes of tubular, blue to purple, two-lipped flowers, each 1–1.5 cm long.

Flowering period	Sept. to Feb.
Distribution	Eastern suburbs, Dandenong Ranges, and widespread throughout the state; also Qld, NSW, ACT, Tas and SA.
Habitat	Open forest and woodland, often in open sites or near rocks.
Notes	An attractive but variable herb which may be prominent after fires.
Similar species	None in the region.
Cultivation	Readily grown in containers or in well-drained soil in an open position.

Family Casuarinaceae

Description

Bushy tree, to 8 m tall, with dark, deeply furrowed bark and slender, upright, green to brown branchlets, which appear as if leafless. Male flowers are borne in fluffy brown spikes, whereas the female flowers are bright red to crimson. The seeds are carried in cylindrical woody cones, 2–2.5 cm long, covered with blunt valves.

Flowering period	March to June.
Distribution	Widespread in the Melbourne region and throughout much of the state; also Qld, NSW and Tas.
Habitat	Open forest and woodland; sometimes dominant and forming almost pure stands.
Notes	During flowering some plants take on a dark brown appearance. Previously well known as *Casuarina littoralis.*
Cultivation	Adaptable to a range of positions in well-drained soil.

15

Family Casuarinaceae

Description

Densely rounded, bushy tree with dark, deeply furrowed bark and slender, drooping, greyish-green branchlets, which appear leafless. Male flowers are borne in slender, yellowish to light brown spikes and the female flowers are reddish. The seeds are carried in cylindrical woody cones, to 4.5 cm long, covered with sharply pointed valves.

Flowering period	March to Dec.
Distribution	Widespread in the Melbourne region and throughout much of the state; also NSW, ACT, Tas and SA.
Habitat	Coastal scrub, heathland, heathy forests, open forest and woodland, often on rocky sites and sometimes in thickets.
Notes	During flowering, the plants take on a light brown to yellowish appearance. Previously well known as *Casuarina stricta*.
Cultivation	Easily grown in well-drained soils. Excellent for exposed coastal sites.

16

Family Loranthaceae

Description

Parasitic plant which forms large clumps of pendulous, fleshy stems on the larger branches of eucalypts and wattles. The stems, which are densely covered with yellowish-green to bronze-green, pendulous leaves, to 20 cm x 2 cm, carry drooping clusters of reddish, tubular flowers (some of which are stalkless), which have narrow spreading to recurved segments.

Flowering period	Sporadic all year.
Distribution	Widespread in the Melbourne region and throughout much of the state; also NSW, ACT and SA.
Habitat	Open forest and woodland.
Notes	Heavier infestations of this natural parasite occur on isolated trees, rather than on trees in forests. The nectar-rich flowers are a food source for birds and the fleshy fruit are eaten by the Mistletoe Bird, which is the major dispersal agent for the plant.
Similar species	*A. miquelii* which grows mainly on box eucalypts and red gums, has clusters of stalked flowers.
Cultivation	By attaching seeds to a suitable branch.

Family Anthericaceae

Description

Lily-like herb with fleshy roots, and a basal rosette of pale green leaves, to 40 cm x 0.5 cm, which can be flat on the ground or erect. Wiry, branched stems, to 1 m tall, carry mauve or violet flowers, about 3 cm across, which have a delightful, chocolate-like fragrance. Each flower, which is borne singly at a node, has six spreading segments and prominent dark anthers.

Flowering period	Mainly Sept. to Dec., but often also sporadic in the autumn.
Distribution	Widespread in the region and throughout the state; also NSW, ACT, Tas and SA.
Habitat	Grassland, open forest and woodland.
Notes	A familiar species which is well known for its chocolate-scented flowers. Each flower lasts one day. Formerly well-known as *Dichopogon strictus.*
Similar species	*A. fimbriatum* has its flowers in clusters, rather than singly at each node.
Cultivation	Readily grown in a container or well-drained soil in sun or semi-shade.

Family Aspleniaceae

Description

Small ground fern forming spreading or sprawling patches, with creeping rhizomes and slender, bright green, once-divided fronds, which have small, wedge-shaped or fan-shaped leaflets, and a long drawn-out tip, on which plantlets can develop.

Distribution	Widespread in the Melbourne region and throughout much of the state; also Qld, NSW, ACT, Tas and NZ.
Habitat	Clefts and crevices of rocks, boulders and cliff faces, and in shallow soil over rock plates; less commonly among shrubs and grasses in open forest and woodland; sometimes epiphytic on tree fern trunks and logs.
Notes	The fresh appearance of this fern belies its tolerance of dryness. Plantlets take root and grow while still attached to the parent frond.
Similar species	None in the region.
Cultivation	An excellent fern for terrariums and hanging containers but also in the garden in well-drained soil in filtered sun.

Family Aspleniaceae

Description

Fern, either growing in the ground, or on tree fern trunks and tree trunks, with an erect, fleshy rootstock and erect to arching, finely divided, lacy, dark green fronds, to 80 cm x 30 cm. Each frond, which is divided two or three times into narrow segments, bears numerous plantlets towards the tip.

Distribution	Dandenong Ranges, and widespread in moist areas of the state; also Qld, NSW, ACT, Tas and NZ.
Habitat	Moist to wet gullies and streambanks in tall forest.
Notes	The tiny plantlets take root as the older fronds decay. Also known as *A. bulbiferum* subsp. *gracillimum*.
Similar species	None in the region.
Cultivation	An attractive container plant or in a sheltered situation in well-drained moisture retentive soils.

Family Epacridaceae

Description

Shrub forming a dense prostrate, spreading mat of crowded, bluish-green, prickly leaves, to about 1 cm long. Bright red, tubular flowers, about 1.5 cm long, appear to erupt from the leaves, but can also be partly hidden within the dense growth. These are followed by pale green or whitish, globular fruit, 0.6–1 cm across.

Flowering period	May to Sept.
Distribution	Widespread in the Melbourne region and throughout lowland areas of the state; also NSW, ACT, Tas, SA and WA.
Habitat	Grassy areas of open forest, woodland and heathland.
Notes	The fruit have a thin layer of sweetish edible flesh.
Similar species	None in the region.
Cultivation	Requires excellent drainage and free air movement.

Family Chenopodiaceae

Description

Sprawling to erect, bushy shrub, to about 2 m tall, with brittle branches densely clothed with silvery grey, somewhat fleshy leaves, to 8 cm x 1.5 cm. Relatively inconspicuous, unisexual flowers, reddish male and cream female, usually occur on separate plants. The small corky fruit are triangular in shape.

Flowering period	Sept. to March.
Distribution	Coastal bayside suburbs and coastal areas of the state; occurs in all states.
Habitat	Coastal sand just above high tide mark and stabilised coastal dunes.
Notes	An important species for stabilising coastal habitats. The leaves are edible after cooking.
Similar species	None in the region.
Cultivation	Excellent for coastal districts, but also adaptable to other regions in well-drained soils.

Family Poaceae

Description

Perennial grass which forms dense, rounded tussocks, to 1 m tall, which often have a bronzed or brownish appearance. Each tussock consists of stiff, hairless, cylindrical, prickly leaves and narrow flower-heads about as long as the leaves. Each pale-coloured flower has a long, twice-bent bristle or awn.

Flowering period	Oct. to March.
Distribution	Eastern and western bayside suburbs, and coastal districts of the state; also NSW, Tas and SA.
Habitat	Coastal dunes, coastal headlands, coastal cliffs and saltmarshes.
Notes	A tough, hardy grass which withstands considerable exposure to buffeting, salt-laden winds. Previously known as *Stipa stipoides*. *S. teretifolia* is a synonym.
Similar species	Many other species of Stipa occur in the Melbourne region but none have cylindrical and prickly leaves.
Cultivation	Requires excellent drainage. Best in coastal districts.

Family Verbenaceae

Description

Shrub or tree, to 8 m tall, with a dense mounded habit and unusual, oxygen-collecting, root-like structures which grow erect from the mud. Flattish stems are densely clothed with thick, leathery, dark green, shiny leaves, to 6 cm x 2.5 cm, which have a prominent yellowish midrib. Small, yellow, fragrant flowers, which are borne in axillary clusters, are followed by leathery, yellowish fruit.

Flowering period	Mainly Jan. to March, but also sporadic at other times.
Distribution	Western bayside suburbs and Westernport Bay; also Qld, NSW, SA, WA and NZ.
Habitat	Coastal estuaries, mud flats and salt marshes.
Notes	An important species for the stabilisation of muddy coastal sites. The fruit was eaten by the Aborigines after roasting.
Similar species	None in the region.
Cultivation	Grown by beach protection authorities. The seed germinates on the bush before falling.

Family Proteaceae

Description

Large, erect to spreading tree, to 20 m tall, with light grey, rough bark and tough, dark green leaves, to 15 cm x 3 cm, which are conspicuously silvery beneath. The leaves may be smooth or have prominently toothed margins. The pale yellow flowers, borne in showy, cylindrical spikes to 15 cm x 6 cm, are followed by grey, woody cones.

Flowering period	Mainly Feb. to Sept., but also sporadic at other times.
Distribution	Eastern and south-eastern bayside suburbs and mainly in eastern coastal areas of the state; also Qld and NSW.
Habitat	Coastal scrub, heathy forest and coastal banksia woodland.
Notes	An important species for soil stabilisation in coastal districts and as a nectar source for birds and possums.
Similar species	None in the region.
Cultivation	An excellent tree for exposed coastal situations, but can also be grown in gardens well away from the coast.

25

Family Proteaceae

Description

This species is extremely variable in growth habit, ranging from a straggly shrub to a dense, small tree to 10 m tall. Its stiff leaves, to 10 cm x 1 cm, are prominently silvery white beneath and the pale yellow to bright yellow flowers, borne in dense spikes, to 10 cm x 5 cm, are followed by grey, bristly cones.

Flowering period	Mainly Feb. to Sept., but also sporadic at other times.
Distribution	Widespread in the Melbourne region and throughout much of the state; also NSW, ACT, Tas and SA.
Habitat	Occurs in many types of habitat including coastal scrub, heathland, heathy forest, open forest and woodland.
Notes	Although the growth form of the plants is variable, the flowers are relatively uniform. Some plants have attractive, brown, furry new growth.
Cultivation	Readily grown and tolerant of a range of positions in moisture-retentive to well-drained soils.

Family Proteaceae   Hairpin Banksia

Description

Dense spreading shrub, to 4 m tall, with stiff, narrow, dark green, toothed leaves, to 10 cm x 0.7 cm, and erect, dense spikes of yellow to golden flowers, to 20 cm x 8 cm, with prominent, hooked, black styles. The flower-spikes are followed by narrow, woody cones covered with the grey remnants of dead flowers.

Flowering period	Feb. to July.
Distribution	Eastern suburbs, Dandenong Ranges, and mainly in eastern areas of the state; also Qld and NSW.
Habitat	Moist sheltered slopes in open forest and woodland, often near streams.
Notes	Plants of this variety lack a lignotuber and are killed by fire. The flowers are highly attractive to nectar-feeding birds. Also known as Golden Candlesticks.
Cultivation	An excellent garden plant requiring well-drained soils.

Family Asteraceae

Description

Shrub or tree, to 7 m tall, with grey, fissured bark and widely
spreading branches clothed towards the tips with large, soft, leaves,
to 18 cm x 4 cm, which are dark green and shiny above and
woolly-white beneath. Yellow, daisy-like flower-heads are carried
in loose, woolly panicles from the upper axils.

Flowering period	Oct. to Jan.
Distribution	Hilly eastern suburbs, Dandenong Ranges, and in southern areas of the state; also NSW and ACT.
Habitat	Cool wet forests and moist gullies in tall forests.
Notes	This species often grows in extensive stands.
Similar species	Previously confused with *B. salicina* which is restricted to Tas.
Cultivation	Generally intolerant of dryness and can be difficult to grow. Best with regular mulching and watering.

Family Pittosporaceae

Description

Bushy climber with slender stems which climb vigorously, to 3 or 4 m tall, and often twine back on themselves to form a tangle. The stems are clothed with narrow, dark green leaves, to 4 cm x 0.5 cm, and masses of pendent, tubular flowers, to 3 cm long, are carried on slender stalks from the upper axils. The flowers, which are yellowish green with purple markings, are followed by oblong, fleshy, shiny, purple berries, about 2 cm x 1 cm.

Flowering period	Aug. to Jan.
Distribution	Occurs in a few eastern suburbs, Dandenong Ranges, and in southern parts of the state; also NSW, ACT and Tas.
Habitat	Moist slopes and gullies in tall forest.
Notes	A showy species with colourful flowers and fruit. The nectar-rich flowers are attractive to birds.
Similar species	*B. scandens* has hairy new growth and green fruit.
Cultivation	Readily grown in a moist sheltered position.

Family Blechnaceae

Description

Ground fern forming clumps or spreading patches, with an erect, thick, woody, black rootstock and straight to arching, light green, once-divided fronds, usually about 1 m long. The leaflets are attached to the main stalk by broad bases. New growth is often yellowish. Fertile and sterile fronds are similar in shape and size.

Distribution	Eastern suburbs, Dandenong Ranges, and in eastern areas of the state; also Qld, NSW, ACT and Tas.
Habitat	Margins of moist forests and streambanks, but also often on slopes in drier open forest.
Notes	Plants reproduce by underground rhizomes and can form thickets, especially on moist sites.
Similar species	*B. nudum* has sterile fronds with broad leaflets and fertile fronds with narrow leaflets.
Cultivation	A very adaptable fern which will tolerate a wide range of soils and positions.

Family Blechnaceae

Description

Ground fern with a short erect rootstock and distinctly different sterile and fertile fronds. The light green sterile fronds, to about 75 cm long, which have short, round to oblong leaflets, spread widely in a nearly prostrate rosette, whereas the dark brown or blackish fertile fronds, with narrow leaflets, are mostly held erect in the centre of the plant.

Distribution	Eastern suburbs, Dandenong Ranges, and in eastern areas of the state; also NSW, Tas and NZ.
Habitat	Moist to wet sheltered sites in forests and in streamside vegetation.
Notes	Plants reproduce by underground rhizomes and can form localised colonies.
Similar species	None in the region.
Cultivation	A very attractive fern which requires shade, humidity and regular watering.

Family Blechnaceae

Description

Ground fern with an erect rootstock, which may develop into a short, fibrous, black trunk, and with distinctly different sterile and fertile fronds. The dark green sterile fronds, to about 90 cm long, develop in a spreading rosette and have numerous, relatively narrow leaflets at right angles to the main stalk. The erect, blackish, fertile fronds are usually shorter than the sterile fronds and have very narrow leaflets held at a steeper angle.

Distribution	Eastern suburbs, Dandenong Ranges, and throughout much of the state; also Qld, NSW, ACT, Tas and SA.
Habitat	Moist to wet sites along streambanks and in sheltered forests; occasionally in exposed sites, especially in wet soils.
Notes	Plants reproduce vigorously from underground rhizomes and may form dense colonies devoid of other plants.
Similar species	*B. cartilagineum* has similar sterile and fertile fronds.
Cultivation	Readily grown in moist soils, but grows best in sheltered situations.

Family Blechnaceae

Description

Ground fern, forming spreading thickets by underground rhizomes, with distinctly different sterile and fertile fronds. The sterile fronds, to about 1 m long, are dark green with broad, coarse leaflets on a short basal stalk. By contrast the fertile fronds are usually taller than the sterile fronds and have much narrower, drawn-out segments.

Distribution	Eastern suburbs, Dandenong Ranges, and throughout much of the state; also Qld, NSW, ACT, Tas and SA.
Habitat	Moist to wet sites along streambanks and in shady forests; occasionally in exposed sites, especially in wet soils.
Notes	Plants usually grow in dense spreading thickets resulting from the prolific production of rhizomes. New fronds have attractive bronze, pink or reddish tones.
Similar species	None in the region.
Cultivation	Readily grown in a sheltered position in moist soil.

Family Asteraceae

Description

Small perennial daisy with a tuft of broad, dark green, spoon-shaped leaves, to 15 cm x 3 cm, which spread in a prostrate rosette. White or blue, typical daisy flower-heads, about 4 cm across, are borne on slender, erect, unbranched stems held well above the rosette of leaves.

Flowering period	Aug. to March.
Distribution	Northern and eastern suburbs, and widespread in cooler parts of the state; also NSW, ACT and Tas.
Habitat	Grassland and grassy woodland.
Notes	May grow as individuals or in small loose groups.
Similar species	None in the region.
Cultivation	Readily adapts to cultivation in containers.

Family Asteraceae

Description

Slender daisy forming sparse clumps, to 30 cm tall, and spreading a similar distance. Most parts are clothed with woolly hairs and the wiry stems are sparsely clothed with narrow, toothed leaves, to about 3 cm long. The typical daisy flower-heads, about 1.5 cm across, with bright yellow centres have spreading white rays. The margins of the seed have distinct lobes.

Flowering period	Oct. to Dec.
Distribution	Western suburbs and western parts of the state; also Qld, NSW and SA.
Habitat	Grassland, grassy forest and woodland.
Notes	*B. heterodonta* is a synonym.
Similar species	*B. ciliaris* has much longer lobes on the leaves.
Cultivation	Requires well-drained soil in a sunny position.

Family Asteraceae

Description

Small perennial daisy which forms compact clumps, to about
30 cm tall, sometimes spreading by slender underground rhizomes.
The dark green, soft, deeply divided leaves form a backdrop to the
typical daisy flower-heads, which can be up to 2 cm across. These
are held on slender stalks above the foliage and can be white, pink,
mauve or blue.

Flowering period	Mainly Aug. to March, but also sporadic at other times.
Distribution	Northern and eastern suburbs and over much of the state; also Qld and NSW.
Habitat	Often among rocks in semi-sheltered sites in open forest and woodland.
Notes	This species is variable in growth habit, leaf division and flower colour.
Similar species	None in the region.
Cultivation	Very popular and readily grown in the garden or in containers.

Family Asteraceae

Description

Annual or short-lived perennial with erect, branched stems, to about 1 m tall, clothed with soft, bright green leaves, to 12 cm x 2 cm. The lowest leaves are largest and they reduce in size up the stems. Bright yellow to golden, typical daisy flower-heads, to 4 cm or more across, with stiff, shiny, papery segments, produce an impressive floral display.

Flowering period	Mainly Oct. to April, but also sporadic at other times.
Distribution	Eastern suburbs and throughout the state; occurs in all states.
Habitat	Heathland, heathy forests and drier forests, often on disturbed sites.
Notes	This species is variable in growth habit and size of the flower-heads. Previously well known as *Helichrysum bracteatum*.
Similar species	*B. viscosa* has much narrower, sticky leaves.
Cultivation	Easily grown in a sunny position in well-drained soil.

Family Goodeniaceae

Description

Perennial herb with a thickened rootstock and basal rosette of silky-haired, light green, spoon-shaped leaves, to 10 cm x 1.5 cm. Tiny, bright blue flowers are borne in dense, button-like or pincushion-like heads, 2–3 cm across, on the end of slender, leafless stems to 50 cm tall.

Flowering period	Oct. to Jan., but also sporadic at other times.
Distribution	Widespread in the Melbourne region and throughout much of the state; occurs in all states.
Habitat	Grassy areas in open forest, woodland and grassland.
Notes	A showy species which superficially resembles a daisy.
Similar species	None in the region.
Cultivation	Excellent in containers, but best treated as an annual in the garden, as plants are often short-lived.

Family Asphodelaceae

Description

Perennial, bulbous, lily-like herb with a dense tussock of upright, bright green, rush-like, fleshy leaves, to about 35 cm long, and erect flower-spikes, to 70 cm tall, with crowded buds and flowers. The bright yellow flowers, to 2 cm across, open widely with six, spreading, starry segments and protruding stamens which carry a prominent brush of hairs immediately below the anther.

Distribution	Widespread in the Melbourne region and throughout much of the state; also Qld, NSW, ACT, Tas and SA.
Habitat	Moist grassland and moist to wet grassy areas in open forest and woodland.
Notes	This species produces impressive floral displays, especially in wet years. Each flower lasts a single day.
Similar species	None in the region.
Cultivation	Readily grown in containers or moisture-retentive soils in the garden.

Family Colchicaceae

Description

Perennial, bulbous, lily-like herb with a loose basal rosette consisting of a few, narrow, shiny green, basal leaves, to 20 cm long. Wiry flower-stems, to 40 cm tall, are topped with a cluster of white, starry, scented flowers. Each flower, about 2.5 cm across, has six spreading segments, six dark anthers and a prominent, triangular, central ovary.

Flowering period	Sept. to Dec.
Distribution	Widespread in the Melbourne region and throughout much of the state; also Qld, NSW, ACT, Vic, Tas, SA and WA.
Habitat	Grassland, heathland, heathy forest, grassy forest, open forest and woodland.
Notes	Often locally common and producing impressive floral displays, especially in wet years and after fires. The flowers are long-lasting.
Similar species	None in the region.
Cultivation	There has been little success in cultivating this plant.

Family Pittosporaceae

Description

Dense, prickly shrub, to 6 m tall, with black spines along the branches and narrow, oblong to spoon-shaped, shiny, dark green leaves, to 2.5 cm x 1 cm. Small, fragrant, creamy white flowers, borne in large terminal clusters, are followed by clusters of papery, brown fruit.

Flowering period	Dec. to March.
Distribution	Mainly northern and eastern suburbs, and throughout the state; also Qld, NSW, ACT, Tas, SA and WA.
Habitat	Open forest, woodland and streamside vegetation.
Notes	Although prickly, this species is well-known for its massed floral displays. The flowers are pleasantly scented and attractive to butterflies.
Similar species	None in the region.
Cultivation	Readily grown in a range of positions in well-drained soil.

41

Family Anthericaceae

Description

Perennial, tuberous, lily-like herb with a dense, crowded tussock of dark green leaves, to 30 cm long, and erect, wiry, flower-stems longer than the leaves. Starry, pale blue to dark blue flowers, 1–1.5 cm across, can produce a showy display. Each flower has six spreading segments which twist in a tight spiral once the flower has withered.

Flowering period	Sept. to Feb.
Distribution	Widespread in the Melbourne region and throughout much of the state; also NSW, ACT, Tas and SA.
Habitat	Grassland, grassy forests, open forest and woodland.
Notes	A variable species with plants from the western suburbs being more robust with larger, darker flowers. Each flower lasts a single day.
Similar species	None in the region.
Cultivation	Readily grown in a container or among rocks in the garden.

Family Orchidaceae

Description

Small terrestrial orchid with a bright green, hairy, basal leaf, to
7 cm x 0.4 cm, which is usually flat on the ground, and a single,
bright blue flower, about 2.5 cm across, borne on a thin wiry stem.
The flower has an erect or recurved rear sepal and four other
segments spreading forwards like the fingers of a hand. The lip has
dark blue bars and two rows of yellow calli.

Flowering period	July to Sept.
Distribution	Northern and eastern suburbs and scattered in the state; also Qld, NSW and ACT.
Habitat	Open forest, heathy forest and heathland.
Notes	A small colourful orchid which is often one of the first species to flower in spring.
Similar species	*C. deformis* has densely crowded blue calli on the lip.
Cultivation	There has been little success in cultivating this plant.

Family Orchidaceae

Description

Small terrestrial orchid with a dark green, erect, hairy basal leaf, to 15 cm x 0.4 cm, and one to four flowers borne on a thin wiry stem, to 25 cm tall. The flowers, which are 2–3 cm across, are whitish to pink, with an erect rear sepal and four other segments spreading forwards like the fingers of a hand. The lip has dark red bars and two rows of yellow calli.

Flowering period	Aug. to Oct.
Distribution	Northern and eastern suburbs, Dandenong Ranges, and throughout much of the state; also Qld, NSW, ACT, Tas and SA.
Habitat	Open forest, heathy forest and heathland.
Notes	A variable species which is often locally common. Plants usually have pink flowers, but in some districts the flowers may be white or greenish.
Similar species	*C. catenata* has larger white flowers and the lip has no red bars; *C. fuscata* has even thinner flower-stems and a single flower.
Cultivation	There has been little success in cultivating this plant.

Family Orchidaceae

Description

Small terrestrial orchid with a dark green, erect, hairy basal leaf, to 20 cm x 0.6 cm, and one to three flowers borne on a thin, wiry stem, to 60 cm tall. The flowers, which are 2.5–3 cm across, are bright pink with conspicuous black calli on the lip. The rear sepal is strongly incurved forwards and four other segments spread widely like the fingers of a hand. The lip is mostly covered by a mass of black calli.

Flowering period	Oct. to Dec.
Distribution	Eastern suburbs, Dandenong Ranges, and in eastern parts of the state; also NSW, ACT, Tas and SA.
Habitat	Among grass tussocks and shrubs in open forest, heathy forest and heathland.
Notes	A very distinctive, late-flowering species which is now uncommon.
Similar species	None in the region.
Cultivation	There has been little success in cultivating this plant.

Family Orchidaceae

Description

Small terrestrial orchid with a thin, bright green, erect, hairy basal leaf, to 10 cm x 0.5 cm, and a single bright blue flower on a slender wiry stem, to 15 cm tall. The flowers, which are 3–4 cm across, have an erect or recurved rear sepal and four other segments spreading forwards like the fingers of a hand. The lip has blue calli in densely crowded rows.

Flowering period	Aug. to Oct.
Distribution	Eastern and south-eastern suburbs, and mainly in western parts of the state; also NSW, Tas, SA and WA.
Habitat	Open forest, heathy forest, heathland and coastal scrub.
Notes	A colourful orchid which is among the first species to flower in spring.
Similar species	*C. caerulea* has two rows of yellow calli on the lip.
Cultivation	There has been little success in cultivating this plant.

Family Orchidaceae

Description

Small terrestrial orchid with a broad, bright green, hairy basal leaf, to 20 cm x 3 cm, which is usually flat on the ground. Bright pink flowers, 3–3.5 cm across, are borne one to four on the end of a wiry stalk, to 40 cm tall. Each has an erect rear sepal and four other segments spreading forwards like the fingers of a hand. The lip has prominent marginal teeth and a central group of thick yellow calli.

Flowering period	Sept. to Oct.
Distribution	Western and eastern bayside suburbs, and in southern parts of the state; also Tas and SA.
Habitat	Coastal scrub, tea-tree heath and heathland.
Notes	This species grows in dense colonies with a very low proportion of the plants flowering.
Cultivation	There has been little success in cultivating this plant.

Family Orchidaceae

Description

Terrestrial orchid with a conspicuously hairy basal leaf, to 12 cm x 1 cm, and one to three spider-like flowers borne on a wiry stalk, to 80 cm tall. Each flower, which can have segments to 6 cm long, is greenish-cream to cream with reddish tips, and has a red lip which coils back at the apex and is fringed with short red teeth.

Flowering period	Aug. to Oct.
Distribution	Eastern suburbs, Dandenong Ranges and a couple of sites in the west of the state.
Habitat	Open forest and woodland.
Notes	In hot weather the flowers exude an unusual smell, like that of hot metal.
Similar species	*C. venusta* has larger white flowers and the lip is fringed with long teeth; *C. rosella* has bright pink flowers.
Cultivation	There has been little success in cultivating this plant.

Family Orchidaceae

Description

Terrestrial orchid with a conspicuously hairy basal leaf, to 8 cm x
0.9 cm, and a single, bright pink, spider-like flower, to 6 cm across,
borne on a wiry stalk, to 17 cm tall. Each flower, which can have
segments to 5 cm long, has a pink to red lip which coils back at the
apex and is fringed with short red teeth.

Flowering period	Aug. and Sept.
Distribution	North-eastern suburbs and a few sites in western parts of the state.
Habitat	Open forest and box eucalypt woodland.
Notes	This species, once locally common in the Eltham district, *has been reduced to great rarity* by urbanisation and is now considered endangered.
Similar species	*C. oenochila* is taller growing with greenish to cream flowers and a red lip.
Cultivation	There has been very little success in cultivating this plant.

Family Orchidaceae

Description

Terrestrial orchid with a conspicuously hairy basal leaf, to 15 cm x 2 cm, and one or two large spider-like flowers, to 10 cm across, on a stem to 50 cm tall. These are greenish with a crimson stripe and have a large green lip with a maroon apex and maroon calli. Each flower can have segments to 8 cm long. The lip has deeply fringed margins and is delicately hinged at the base.

Flowering period	Oct. to Nov.
Distribution	Relatively widespread in the Melbourne region and throughout much of the state; also SA.
Habitat	Open forest, woodland, heathy forest and heathland.
Notes	An exciting orchid with large flowers and a very unusual lip which is mobile, even trembling in a slight breeze.
Similar species	*C. phaeoclavia* has smaller flowers with brownish clubs on the end of the sepals.
Cultivation	There has been little success in cultivating this plant.

Family Orchidaceae

Description

Terrestrial orchid with a reddish-pink, spotted basal leaf, to 12 cm x 0.8 cm, and a thin, wiry flower-stem, to 40 cm tall, carrying one to five flowers shaped remarkably like flying ducks. Each reddish-brown flower has the narrow segments recurved or swept back and the lip, which resembles a duck's head complete with a broad beak, sits atop a thick springy band.

Flowering period	Oct. to Dec.
Distribution	Eastern suburbs and throughout southern parts of the state; also Qld, NSW, Tas and SA.
Habitat	Open forest, woodland, heathy forest and heathland.
Notes	A remarkable orchid which usually grows in small colonies. The lip is sensitive to touch and curls inwards against the sexual organs when triggered.
Similar species	None in the region.
Cultivation	There has been little success in cultivating this plant.

Family Myrtaceae

Description

Shrub or tree, to 10 m tall, with an open to dense appearance, silvery, silky, new growth and spreading to pendulous branches. Leathery, dark green leaves, to 10 cm x 0.8 cm, taper to each end. Cream or pink bottlebrush-like flowers, to 8 cm x 3 cm, which are borne near the end of the branches, are followed by clusters of small, woody fruit.

Flowering period	Nov. to May.
Distribution	Scattered throughout the Melbourne region and in many areas in the state; also NSW, ACT, Tas and SA.
Habitat	Streamside vegetation, usually close to the water.
Notes	The nectar-rich flowers are very attractive to birds. Commonly this species has cream flowers but a pink selection is frequently cultivated. Previously well known as *C. paludosus*.
Similar species	None in the region.
Cultivation	Readily grown in a variety of soils and positions. Requires regular pruning.

Family Cupressaceae

Description

Bushy or spreading tree, to 20 m tall, with a straight trunk covered with grey, fissured bark. New growth is bluish-green and the twiggy branchlets are covered with tiny, greyish-green, scale-like leaves. Tiny brown male cones arise at the end of the branchlets, whereas the much larger, rounded, grey female cones, to 2.5 cm across, are borne well in from the tips.

Flowering period	Sept. to Nov.
Distribution	Occurs in a few north-western suburbs and mainly in northern parts of the state; also Qld, NSW, SA, WA and NT.
Habitat	Grassland and woodland.
Notes	The numbers of this species have been greatly reduced in the region due to agricultural pursuits and also from rabbits eating the seedlings.
Similar species	None in the region.
Cultivation	Readily grown in a sunny position in well-drained soils.

Family Orchidaceae

Description

Terrestrial orchid with an erect, three-cornered, green leaf, to 40 cm x 0.8 cm, which is often reddish at the base. A sturdy flower-stem, to 45 cm tall, carries one to nine flowers, 2–3 cm across. These have green or reddish segments with purplish stripes and a large lip covered with purple glands and hairs.

Flowering period	Oct. and Nov.
Distribution	Eastern suburbs, Dandenong Ranges, and throughout much of the state; also Qld, NSW, ACT, Tas and SA.
Habitat	Open forest, woodland, heathy forest and heathland.
Notes	The central part of the flowers of this orchid resemble a bearded face.
Similar species	*C. paludosus* has reddish hairs on the lip.
Cultivation	There has been little success in cultivating this plant.

Family Culcitaceae

Description

Coarse ground fern forming widely spreading clumps, with vigorous, coarse rhizomes and broad, green to yellowish-green fronds, to 150 cm x 80 cm. Each frond has a dark brown, hairy stalk and broadly triangular, hairy blade which is divided three or four times. Sterile and fertile fronds are of similar shape and size. Spores are borne in small round clusters covered by a small flap.

Distribution	Northern and eastern suburbs, Dandenong Ranges, and southern areas of the state; also Qld, NSW, ACT and Tas.
Habitat	Open forest and woodland, often colonising disturbed sites such as road verges.
Notes	A common fern which frequently grows in extensive dense patches. On exposed sites the fronds are usually yellowish, whereas in shady situations they are green. Previously well known as *Culcita dubia*.
Similar species	None in the region.
Cultivation	Very easily grown in a range of positions in well-drained soil.

Family Asteraceae

Description

Dwarf perennial shrub which forms a tangled, mounded clump of wiry stems and bright green foliage comprised of wedge-shaped hairy leaves, to 6 cm x 1 cm, with toothed or lobed margins. Bright yellow flowers, borne in globular, burr-like heads, about 1 cm across, contrast pleasantly with the foliage.

Flowering period	Mainly Aug. to March, but also sporadic at other times.
Distribution	North-western suburbs and scattered disjunctly in the state; also Qld, NSW, ACT, SA and WA.
Habitat	Open forest and woodland, sometimes on disturbed sites and road verges.
Notes	The spiny fruit (burrs) are very prickly and catch in woolly clothing and animal fur.
Similar species	None in the region.
Cultivation	Easily grown in a sunny position in well-drained soil, but the prickly fruit are a drawback.

Family Myrtaceae

Description

Small erect or spreading, often dense shrub, to 2 m tall, with tiny, bright green, aromatic leaves crowded along the branchlets. Satiny-white, starry flowers, carried in densely crowded terminal heads, have stiff petals, a mass of protruding stamens, and fine, thread-like structures on the sepals. The dark red calyx is noticeable throughout the ripening period.

Flowering period	Aug. to Nov.
Distribution	North-western suburbs and throughout the state; also NSW, ACT and SA.
Habitat	Rocky slopes in open forest and near streams.
Notes	Flowering plants bear a profusion of well-displayed showy flowers. The leaves exude a pleasant fragrance when crushed.
Similar species	None in the region.
Cultivation	An attractive shrub which is readily grown in a sunny position in well-drained soil.

Family Aizoaceae

Description

Spreading perennial with prostrate, creeping stems and clusters of fleshy, dark green, succulent leaves, to 10 cm long. Showy mauve to purple flowers, each 4–5 cm across, are borne profusely among the leaves and are followed by globular, purplish, fleshy fruit.

Flowering period Mainly Aug. to Feb., but also sporadic at other times.

Distribution Northern and eastern suburbs and widely scattered in the state, mainly in coastal districts; also Tas, SA and WA.

Habitat Coastal dunes, coastal cliffs, headlands and coastal scrub.

Notes The flowers of this species can have up to 200 petals. The fleshy leaves and fruit were eaten by the Aborigines.

Similar species *C. modestus* has smaller leaves and flowers.

Cultivation Readily grown in a sunny location in well-drained soil.

Family Asteraceae

Description

Bushy shrub, to 4 m tall, usually with a spreading habit, the branchlets appear white due to their dense covering of hairs. Narrow, dark green, somewhat roughened leaves, to 4 cm x 0.3 cm, with recurved margins, smell strongly when crushed. Small, white, flower-heads are borne in large, dense, prominent, flat-topped clusters at the end of the branchlets.

Flowering period	Nov. to March
Distribution	Widespread in the Melbourne region and over much of the state; also NSW, ACT, Tas and SA.
Habitat	Open forest and woodland.
Notes	A hardy species which can colonise disturbed sites.
Similar species	*C. longifolia* has much larger leaves which are sticky when young.
Cultivation	Easily grown in a range of well-drained soils and positions.

Family Asteraceae

Description

Bushy shrub, to 3 m tall, with erect, densely hairy branchlets and narrow, drooping leaves, to 1 cm x 0.1 cm, which have a spicy fragrance when crushed. At flowering time, the whole shrub is covered with long, drooping clusters of shiny brown flower-heads.

Flowering period	Nov. to March.
Distribution	Widespread in the Melbourne region and over much of the state; also NSW, ACT, SA and WA.
Habitat	Open forest and woodland, especially on disturbed sites and embankments.
Notes	A fast-growing species which was apparently widely distributed on the goldfields by Chinese diggers.
Similar species	None in the region.
Cultivation	Readily grown in a range of soil types and positions.

Family Anthericaceae

Description

Small, perennial, tuberous, lily-like herb with a basal rosette of a few narrow, green or bluish spreading leaves, to 15 cm x 1 cm, and wiry, branched flower-stems, to 20 cm tall. The bright blue flowers, 1–1.5 cm across, have six spreading segments which twist in a tight spiral at the end of flowering. Each petal has three dark blue veins. The fruit is lobed.

Flowering period	Aug. to Nov.
Distribution	Widespread in the Melbourne region and over lowland areas of the state, particularly in the west; also Tas, SA and WA.
Habitat	Grassy forests, open forest, woodland and heathland.
Notes	An attractive little species which sometimes occurs in dense patches. Flowering is stimulated by fires. Each flower lasts a single day.
Similar species	None in the region.
Cultivation	Readily grown in a container but difficult to grow in the garden.

Family Sinopteridaceae

Description

Small ground fern forming dense spreading clumps with crowded, bright green fronds, to 40 cm x 15 cm. Each frond has a wiry, brown stalk and triangular nearly hairless blade, which is divided two or three times. Sterile and fertile fronds are similar in size and shape.

Distribution	Widespread in the Melbourne region and lowland areas of the state; also NSW, Vic, Tas, SA and WA.
Habitat	Among rocks in open forest and woodland.
Notes	A very hardy fern, the fronds of which may die over summer but are rapidly replaced following good autumn rains.
Similar species	*C. sieberi* has much narrower fronds.
Cultivation	Susceptible to disturbance and can be very difficult to establish.

Family Orchidaceae

Description

Small terrestrial orchid with two ground-hugging, dark green leaves, to 6 cm x 3 cm, with crinkled margins. A slender, pinkish flower-stem, to 18 cm tall, carries a single, greenish or pinkish flower, to 1.5 cm long, which has an unusual, insect-like lip, covered with variously arranged black calli.

Flowering period	Feb. to May.
Distribution	Eastern suburbs, Dandenong Ranges, and scattered throughout southern parts of the state; also NSW and Tas.
Habitat	Moist areas of open forest, coastal scrub, heathy forest and heathland.
Notes	This species grows in dense, sometimes extensive colonies with only a low proportion of flowering plants. The flowers develop as the leaves emerge through the ground. They are pollinated by male thynnine wasps which attempt to mate with the lip.
Similar species	None in the region.
Cultivation	Grown in containers by specialist growers but can be difficult to maintain.

Family Orchidaceae

Description

Small terrestrial orchid with two ground-hugging, dark green leaves, to 10 cm x 4 cm, with entire margins. A stiffly erect, greenish to purplish flower, about 3.5 cm across, on a stalk to 7 cm tall, has a broad, heart-shaped, delicately hinged lip with a few dark column-like calli towards the base.

Flowering period	Oct. to Dec.
Distribution	Eastern and south-eastern suburbs, Dandenong Ranges, and mainly in eastern parts of the state; also NSW, ACT, Tas and NZ.
Habitat	Moist slopes and gullies in tall forest.
Notes	The flowers of this species resemble a small bird with its beak open waiting to be fed. It is pollinated in a similar way to *C. reflexa,* but by a different species of wasp.
Similar species	None in the region.
Cultivation	There has been very little success in cultivating this plant.

Family Asteraceae

Description

Prostrate or sprawling herb with semi-erect to spreading, hairy stems, covered in soft, densely hairy, silvery-grey, woolly leaves, to 6 cm x 0.8 cm. Each stem ends in a loose to dense cluster of bright yellow, button-like flower-heads, 1–1.5 cm across.

Flowering period	Mainly Sept. to Jan., but also sporadic at other times.
Distribution	Widespread in the Melbourne region and throughout much of the state; occurs in all states.
Habitat	Open sites in heathland, open forest, woodland and grassland; may also colonise disturbed sites.
Notes	An extremely variable species with variation exhibited in growth habit, leaf size, leaf colour and flower colour. Occasional hybrids occur with *C. semipapposum.*
Similar species	*C. semipapposum* has clusters of stems arising from creeping rhizomes.
Cultivation	A rewarding free-flowering species. Requires free drainage in a sunny position.

Family Asteraceae   **Clustered Everlasting**

Description

Perennial herb with creeping rhizomes and clusters of erect stems, to 80 cm tall, which have narrow, grey to green leaves, to 5 cm x 0.5 cm, which are sometimes sticky. Each stem bears a dense terminal cluster of yellow to orange flower-heads, 0.5–1 cm across.

Flowering period	Oct. to March, but also sporadic at other times.
Distribution	Widespread in the Melbourne region and throughout much of the state; occurs in all states.
Habitat	Rocky areas, ridges and slopes in open forest and woodland.
Notes	A variable species with variation exhibited in growth habit, leaf size and colour and the size and colour of the flower-heads. This species sometimes occurs in extensive stands. Occasional hybrids occur with *C. apiculatum.*
Similar species	*C. apiculatum* has sprawling stems and silvery grey leaves.
Cultivation	Easily grown in well-drained soil in a sunny position. Responds well to pruning.

Family Ranunculaceae

Description

Vigorous climber with slender pliant stems and trifoliolate leaves,
with shiny, green, ovate leaflets, to 8 cm x 2.5 cm, sometimes
having toothed margins. Creamy-white, 4-petalled, starry, unisexual
flowers, 5–6 cm across, are borne in well-displayed clusters from
the end of short side growths. On female plants these are followed
by clusters of feathery seed-heads. Leaves of juvenile shoots are
often purplish with silvery markings.

Flowering period	Widespread in the Melbourne region and throughout the state; also Qld, NSW, ACT and Tas.
Habitat	Moist areas of open forest and woodland, often near streams.
Notes	This species can produce striking floral displays. Male and female flowers are borne on separate plants.
Similar species	*C. microphylla* has much smaller leaflets which have entire margins.
Cultivation	A versatile climber which grows well in moist soils.

Family Ranunculaceae

Description

Bushy climber with slender pliant stems and divided leaves bearing small, dull green, oblong leaflets, to 3 cm x 0.6 cm. Greenish-cream, 4-petalled, starry, unisexual flowers, 3–4 cm across, are borne in small clusters. On female plants these are followed by greyish clusters of feathery seed-heads.

Flowering period	July to Nov.
Distribution	Widespread in the Melbourne region and throughout the state; also Qld, NSW, Tas, SA and WA.
Habitat	Heathland, heathy forest, open forest, woodland and streamside vegetation.
Notes	Male and female flowers are borne on separate plants.
Similar species	*C. aristata* has much larger leaflets which often have toothed margins.
Cultivation	Requires excellent drainage in a sunny or semi-shady position.

Family Polygalaceae

Description

Small shrub, to 1.5 m tall, with slender, erect stems clothed with narrow, oblong, dark green leaves, 1–1.5 cm long. Small, bright pink flowers are carried in dense racemes which arise from the apex of each branch, and sometimes also from the upper leaf axils. Each flower has two cupped segments which spread like wings.

Flowering period	Oct. to Feb.
Habitat	Heathland and heathy forest.
Distribution	Occurs in a few eastern suburbs and is widespread in the state; also Qld, NSW, ACT and Tas.
NotesThe	Flowers of this species superficially resemble those of members of the pea family.
Similar species	*C. calymega* is sparsely branched with blue flowers.
Cultivation	Can be grown in containers, but is difficult to establish in the garden.

Family Polygalaceae

Description

Climber with long, slender, twining, nearly leafless stems, these sometimes twisting around themselves to form bushy clumps. Bright blue flowers, about 0.8 cm across, are borne in showy, terminal clusters. Each flower has two spreading wing-like segments.

Flowering period	Aug. to Dec.
Distribution	Northern and eastern suburbs, Dandenong Ranges, and throughout much of the state; also Qld, NSW, ACT, Tas, SA and WA.
Habitat	Open forest, heathy forest and heathland.
Notes	A decorative species which scrambles through surrounding vegetation.
Similar species	None in the region.
Cultivation	There has been little success in cultivating this plant.

Family Rubiaceae

Description

Bushy shrub, to 2 m tall, with broad, dull green, elliptical, pointed leaves, to 7 cm x 1.5 cm, which are rough to the touch. Clusters of tiny, greenish, unisexual flowers are borne in the upper leaf axils, the female flowers being followed by succulent, reddish-orange fruit, to 1 cm long, ripening in autumn.

Flowering period	Aug. to Oct.
Distribution	Dandenong Ranges and widely distributed in cooler parts of the state; also NSW, ACT and Tas.
Habitat	Moist to wet forests and fern gullies.
Notes	Also known as Coffee Berry. The fruit are edible but leave an unpleasant aftertaste.
Similar species	*C. quadrifida* has spiny stems and much smaller, dark green leaves.
Cultivation	Readily grown in a moist sheltered position.

Family Rutaceae

Description

Dense, bushy shrub, to 2 m tall, with a widely spreading habit and an overall greyish appearance due to the colour of the leaves. These are oval to nearly round, thick, and with a paler underside which contrasts with the darker upper surface when the wind blows. Starry, white flowers, to 1.5 cm across, in clusters at the end of branchlets, contrast pleasantly with the foliage.

Flowering period	Mainly Nov. to May, but also sporadic at other times.
Distribution	Eastern bayside suburbs and in coastal districts of the state; also NSW, Tas and SA.
Habitat	Coastal scrub on dunes and headlands.
Notes	An important coastal species which withstands buffeting salt-laden winds. The leaves can apparently be used as a tea substitute.
Similar species	None in the region.
Cultivation	An excellent shrub for coastal districts, but also successfully grown in other areas.

Family Rutaceae

Description

Bushy shrub, to 3 m tall, with slender spreading branches and shiny, dark green, oval to oblong leaves, to 4 cm x 2 cm, which can have wavy margins. Tubular, slender, pale green, bell-like flowers, 2–3 cm long, are carried in the upper axils and on the end of short side shoots.

Flowering period	Mainly April to Aug., but also sporadic at other times.
Distribution	Widespread in the Melbourne region and throughout much of the state; also Qld, NSW and SA.
Habitat	Streamside vegetation and in rocky areas.
Notes	The nectar-rich flowers are highly attractive to birds.
Similar species	*C. reflexa* has dull leaves which are hairy on the upper surface.
Cultivation	Readily grown in well-drained soils in partially protected positions.

Family Rutaceae

Description

Variable, open to dense shrub with a growth habit ranging from sprawling to erect, the slender stems being densely hairy.
Dark green, rough or hairy, dull, oval leaves, to 5 cm x 0.8 cm, are scattered in pairs or groups along the branches. Conspicuous tubular bells, to 2.5 cm long, carried at the end of short branchlets, are commonly light green, but in some areas may be red with yellow tips.

Flowering period	Mainly March to Sept., but also sporadic at other times.
Distribution	Widespread in the Melbourne region and throughout the state; also Qld, NSW, ACT, Tas, SA and WA.
Habitat	Dry to moist sites in open forest, woodland, heathy forest and heathland.
Notes	The flowers are very attractive to nectar-feeding birds.
Similar species	*C. glabra* has shiny leaves which are hairless on the upper surface.
Cultivation	A very desirable garden plant which adapts well to cultivation.

Family Orchidaceae

Description

Tiny terrestrial orchid with a single rounded leaf, to 3 cm x 2 cm, which is dark green on the upper surface and silvery beneath. A single dark purple flower, about 2 cm long, squats near the base of the leaf. This flower has a strongly hooded rear sepal and a broad cupped lip with incurved margins, and a prominent white central patch.

Flowering period	July and Aug.
Distribution	Eastern and south-eastern suburbs, and widespread throughout the state; also NSW, ACT, Tas and SA.
Habitat	Sheltered sites in open forest, coastal scrub, heathy forest and heathland.
Notes	This species grows in dense colonies, with only a very low proportion of plants producing flowers.
Similar species	*C. diemenicus* has larger, erect flowers on a longer stalk.
Cultivation	Grown successfully in containers by specialist growers.

Family Orchidaceae

Description

Tiny terrestrial orchid with a single, rounded to ovate leaf, to 3 cm x 2 cm, which is greyish-green above and reddish-purple beneath. A single dark purple flower, about 1.5 cm long, arises from near the base of the leaf. This flower has a strongly hooded rear sepal and a tubular lip, with a beak-like opening.

Flowering period	June and July.
Distribution	Eastern bayside suburbs and a few other eastern suburbs, but *now very rare;* also in southern parts of the state, NSW, Tas and SA.
Habitat	Moist areas in open forest, heathy forest, coastal scrubs and heathland.
Notes	A distinctive species which usually grows in small colonies.
Similar species	None in the region.
Cultivation	There has been very little success in cultivating this plant.

Family Orchidaceae

Description

Clumping terrestrial orchid with one to a few erect, stalked leaves, to 15 cm x 3 cm, which are dark green to yellowish-green on both surfaces. Stiff, reddish flower-stems, to 80 cm tall, bear unusual reddish-brown flowers, about 3 cm long. These have very narrow segments and a large lip with a blackish lobed callus towards the apex.

Flowering period	Dec. to Feb.
Distribution	Eastern suburbs, Dandenong Ranges, and scattered in southern parts of the state; also Qld, NSW, Tas and SA.
Habitat	Usually moist to wet sites in open forest, heathy forest and heathland, but sometimes also in relatively dry places.
Notes	The flowers of this species are pollinated by the males of a species of ichneumon wasp as they attempt to mate with the lip.
Similar species	None in the region.
Cultivation	Grown successfully in containers by specialist growers.

Family Cyatheaceae

Description

Tree fern with an erect woody or fibrous trunk, to about 10 m x 40 cm, the upper parts covered with old roughened or prickly frond bases, and topped with a graceful crown of widely spreading, dark green fronds. Frond stalks are rough and scaly. On large plants the fronds grow to 3.5 m x 1 m. New fronds are often produced in attractive, pale growth flushes which contrast with the darker mature fronds.

Distribution	Mainly eastern suburbs, Dandenong Ranges, and primarily in eastern parts of the state; also Qld, NSW, ACT and Tas.
Habitat	Slopes and valleys in moist forests, especially near streams; often colonises suitable sites in drier habitats, including disturbed areas.
Notes	Plants tolerate considerable sun if the roots are moist to wet.
Similar species	*C. cunninghamii* has a very slender trunk with a crown of smaller fronds, to about 2 m x 1 m.
Cultivation	A very hardy tree fern which has proved to be adaptable and sun tolerant if watered regularly.

Family Cyperaceae

Description

Annual herb with a loose basal tuft of dark green leaves, to
40 cm x 0.3 cm, and erect flowering stems, to 40 cm tall, which
are triangular in cross-section. Each stem ends in a tuft of leaf-like
bracts and dense clusters of tiny green flowers with dark reddish-
brown margins.

Flowering period	Nov. to May.
Distribution	Northern and eastern suburbs and widespread throughout the state; also Qld, NSW, SA, WA and many overseas countries.
Habitat	Moist to wet depressions, drains and swamps.
Notes	The tiny flowers are attractive when viewed up close.
Similar species	*C. lucidus* grows to 1.5 m tall and has longer flower spikes.
Cultivation	Readily grown in a boggy site. May become weedy.

Family Orchidaceae

Description

Small terrestrial orchid with a ground-hugging, heart-shaped leaf, to 4 cm x 3 cm, which is greyish-green above, with paler veins, and silvery beneath. Pale pinkish-brown flowers, with narrow segments, to 1 cm long, are carried on thin racemes, to 15 cm tall. The reddish lip is like a flat, oblong plate with two prominently raised basal glands.

Flowering period	July to Oct.
Distribution	Eastern and south-eastern suburbs, and widespread in the state; also Qld, NSW, ACT, Tas and SA.
Habitat	Moist sites in open forest, heathy forest and heath.
Notes	This species grows in dense colonies with only a low proportion of the plants producing flowers.
Similar species	None in the region.
Cultivation	Grown successfully in containers by orchid specialists.

Family Fabaceae

Description

Prickly spreading shrub, to 1.5 m tall, with slender, stiff, spine-tipped, tangled branches clothed in dark green, heart-shaped leaves, to 2.5 cm x 0.6 cm. Masses of yellow and brown, pea-shaped flowers, in small clusters from the leaf axils, are followed by small, triangular pods.

Flowering period	Aug. to Nov.
Distribution	Northern and eastern suburbs, Dandenong Ranges, and widespread throughout the state; also Qld, NSW, ACT, Tas, WA and SA.
Habitat	Open forest, woodland, streamside vegetation and heathland.
Notes	Although prickly, this species produces attractive floral displays.
Similar species	None in the region.
Cultivation	Requires well-drained soil in a sunny to semi-shady position.

Family Phormiaceae

Description

Robust, perennial, lily-like herb with erect
dark green leaves, to 70 cm x 1.5 cm, in
fans or tufts, with recurved, finely-toothed
margins. Wiry, branched flower-stems, to
1 m tall, carry blue flowers, about 1 cm
across. Each flower has six spreading or
recurved segments, prominent dark brown
anthers, and is followed by shiny, dark blue, fleshy berries.

Flowering period	Aug. to March.
Distribution	Widespread in the Melbourne region and throughout much of the state; also Qld, NSW, ACT, Tas, SA and WA.
Habitat	Grassland, heathland, heathy forests, open forest and woodland.
Notes	This species forms large clumps and spreads by vigorous underground rhizomes.
Similar species	*D. tasmanica* has broader leaves with flat margins and yellow anthers; *D. longifolia* has paler flowers and smooth leaf margins.
Cultivation	A very hardy and adaptable species.

Family Phormiaceae

Description

Robust, perennial, lily-like herb with fans or tufts of erect, dark green leaves, to 1 m x 4 cm, with flat, toothed margins. Wiry, branched flower-stems, to 1.5 m tall, carry blue flowers, about 1 cm across. Each flower has six spreading or recurved segments, prominent yellow anthers, and is followed by bright blue, shiny, fleshy, oblong berries, which ripen in the autumn.

Flowering period	Aug. to Jan.
Distribution	Eastern suburbs, Dandenong Ranges, and widespread in the state; also NSW, ACT and Tas.
Habitat	Cool sheltered slopes and streambanks in moist to wet forests.
Notes	Some plants spread vigorously by rhizomes. Clusters of berries can produce a colourful display.
Similar species	*D. revoluta* has narrower leaves with recurved margins and dark brown anthers; *D. longifolia* has paler flowers and smooth leaf margins.
Cultivation	Readily grown in a sheltered position.

Family Convolvulaceae

Description

Spreading herb with prostrate, creeping, much-branched, thin stems which take root from the nodes. The bright green, hairy, kidney-shaped leaves, to 4 cm wide, on long stalks, are the most conspicuous feature of this species, as the tiny, greenish flowers can only be seen after careful searching.

Flowering period	Sept. to Dec.
Distribution	Widespread in the Melbourne region and throughout much of the state; occurs in all states.
Habitat	Moist situations in most habitats.
Notes	This species, which forms sparse to dense mats, can become extremely vigorous in moist sheltered situations.
Similar species	None in the region.
Cultivation	Easily grown and can become weedy. Sometimes used as a lawn substitute.

Family Dicksoniaceae

Description

Tree fern with an erect, brown, fibrous trunk, to about 12 m x 1 m, the upper parts covered with smooth, woody frond bases and topped with a graceful crown of widely spreading, dark green fronds. Frond stalks are initially hairy but become smooth. On large plants the fronds grow to 4 m x 80 cm. Flushes of new fronds are an interesting feature.

Distribution	Eastern suburbs, Dandenong Ranges, and in southern parts of the state; also Qld, NSW, ACT and Tas.
Habitat	Moist to wet slopes in tall forest and fern gullies.
Notes	In wet humid areas the fibrous trunks of this species are often festooned with ferns and other epiphytes.
Similar species	Commonly confused with *Cyathea australis* which has prickly frond bases and the young parts are covered with scales.
Cultivation	Readily grown in a sheltered position. Needs regular watering even when established.

Family Fabaceae

Description

Shrub, to 2 m tall, which varies in growth habit from spreading to erect, but always with wiry, twiggy branches and small, cylindrical, dark green leaves, to 2.5 cm long. Colourful, pea-shaped flowers, about 1 cm across, usually in combinations of bright yellow and red, are carried in dense terminal clusters, and followed by small, inflated, sparsely hairy pods.

Flowering period	Aug. to Dec.
Distribution	Eastern suburbs and throughout much of the state; also Qld, NSW, Tas and SA.
Habitat	Open forest, woodland, heathy forest and heathland.
Notes	The flowers of this species are much wider than they are long.
Similar species	*D. cinerascens* has greyish green cylindrical leaves.
Cultivation	Requires filtered sun or partial sun and excellent drainage.

Family Orchidaceae

Description

Leafless terrestrial orchid with thick, green to reddish-black flower-stems, to 80 cm tall, arising singly or in groups directly from a subterranean rhizome. The apex of each stem is a loose, open raceme of 15–50 flowers, 2–3 cm across. These are pale rose-pink with darker spots and the tips of the segments curve back. The lip has darker stripes and a central group of mauve hairs.

Flowering period	Dec. to Feb.
Distribution	Northern and eastern suburbs, Dandenong Ranges, and throughout much of the state; also NSW, ACT, Tas and SA.
Habitat	Open forest, woodland and heathy forest.
Notes	A large, colourful, familiar orchid which is prominent in good seasons.
Similar species	None in the region.
Cultivation	Impossible to grow.

Family Orchidaceae

Description

Terrestrial orchid with a basal tuft of five to nine narrow, grass-like, green leaves, to 15 cm x 0.3 cm, and a central flower-stem, to 30 cm tall, which carries one to four semi-nodding flowers, 2–2.5 cm across. These are bright lemon-yellow with a few brown streaks and markings. The lip, which is mostly flat with a central ridge, projects forwards.

Flowering period	Sept. and Oct.
Distribution	Eastern and south-eastern suburbs, and throughout much of the state; also NSW, ACT, Tas and SA.
Habitat	Moist grassy areas in open forest and woodland.
Notes	A distinctive orchid which often grows in loose colonies.
Similar species	None in the region.
Cultivation	Grown successfully in containers by orchid specialists.

88

Family Orchidaceae

Description

Terrestrial orchid with one to three narrow basal leaves, to 20 cm x 0.5 cm, folded along the midline, and a central flower-stem, to 20 cm tall, which carries one to six, strongly scented flowers, 3–3.5 cm across. These are white to pale mauve with purplish markings. Two long, green sepals project stiffly beneath the fan-shaped lip.

Flowering period	Oct. and Nov.
Distribution	Restricted to the basalt plains immediately north and west of Melbourne.
Habitat	Grassland.
Notes	This attractive species is now *on the verge of extinction* due to urbanisation and invasion of its habitat by weeds. The flowers are delightfully fragrant on a warm day.
Similar species	*D. punctata* is taller growing with purple flowers.
Cultivation	Grown successfully in containers by orchid specialists.

Family Orchidaceae

Description

Terrestrial orchid with one to three narrow basal leaves, to 30 cm x
1 cm, folded along the midline, and a central flower-stem, to 45 cm
tall, which carries one to eight flowers, 3.5–5 cm across. These are
yellow with reddish-brown suffusions in a wallflower pattern. The
lip has a prominent projecting central lobe and two spreading side
lobes of similar length.

Flowering period	Sept. to Nov.
Distribution	Eastern suburbs, Dandenong Ranges, and widespread in the state; also NSW, Tas and SA.
Habitat	Open forest, heathy forest, woodland, heathland and grassland.
Notes	This species is somewhat variable in colour with some colonies being pure yellow and others heavily suffused. Flowering is strongly promoted by fire.
Similar species	None in the region.
Cultivation	Grown successfully in containers by orchid specialists.

Family Orchidaceae

Description

Terrestrial orchid with one to three narrow basal leaves, to 30 cm x 0.5 cm, folded along the midline, and a central flower-stem, to 60 cm tall, which carries one to ten flowers, 4–5 cm across. These are purple with a prominent yellow central patch on the base of the lip. Two long, purplish sepals project stiffly beneath the wedge-shaped lip.

Flowering period	Oct. and Nov.
Distribution	Eastern suburbs, where once relatively common but now *on the verge of extinction* in the region; also scattered throughout the state and in NSW, ACT and SA.
Habitat	Grassland and grassy woodland.
Notes	This species grows in colonies, sometimes extensive. It was once a prominent plant on some suburban railway enclosures.
Similar species	*D. fragrantissima* is short with white to pale mauve flowers.
Cultivation	Grown successfully in containers by orchid specialists.

Family Orchidaceae

Description

Terrestrial orchid with one to three narrow basal leaves, to
50 cm x 0.4 cm, folded along the midline, and a central flower-
stem, to 60 cm tall, which carries one to seven flowers, 2.5–3 cm
across. These are bright yellow with bold, dark brown markings,
two large blotches being prominent at the base of the upper sepal.
The projecting lip is folded along the midline.

Flowering period	Oct. to Dec.
Distribution	Eastern suburbs, Dandenong Ranges, and throughout much of the state; also Qld, NSW, ACT, Tas and SA.
Habitat	Open forest, woodland, heathy forest, heathland and grassland.
Notes	An impressive species which grows in loose colonies and is often locally common.
Similar species	None in the region.
Cultivation	Grown successfully in containers by orchid specialists.

Family Droseraceae

Description

Slender climber with a subterranean tuber and weak, thin stems which scramble through surrounding vegetation, sometimes reaching a height of more than 60 cm. The stems bear small, rounded, yellowish, cupped leaves covered with specialised sticky hairs, which can trap small insects. White to pink, fragrant flowers, 2–2.5 cm across, are carried in small groups at the top of the stem.

Flowering period	July to Dec.
Distribution	Northern and eastern suburbs, and throughout the state; also NSW, Tas, SA and WA.
Habitat	Open forest, woodland, heathy forest and heathland.
Notes	The stems of this species thread through surrounding shrubs and are supported by the sticky leaves. Previously known as *D. planchonii*.
Similar species	None in the region.
Cultivation	Can be successfully cultivated in containers by specialist growers.

Family Droseraceae

Description

Slender herb with a subterranean tuber and stiffly erect, thin stems, to 50 cm tall. These have a small basal rosette and numerous pale green to yellowish stem leaves, which consist of a slender stalk and rounded blade covered with glistening sticky hairs that can trap small insects. White or pale pink flowers, 1–1.5 cm across, are carried in small groups at the top of the stem.

Flowering period	Aug. to Dec.
Distribution	Widespread in the Melbourne region and throughout the state; also Qld, NSW, ACT, Tas, SA and WA.
Habitat	Grassland and grassy or mossy areas in open forest, woodland and heathland.
Notes	This species sometimes grows in dense colonies.
Similar species	*D. peltata ssp. auriculata* lacks a basal rosette and has darker green leaves.
Cultivation	A container plant for specialised growers.

Family Droseraceae

Description

Small perennial herb with a reddish subterranean tuber and a ground-hugging rosette of green, bronze or reddish, spoon-shaped leaves, to 2.5 cm x 1 cm, covered with glistening sticky hairs. Large, scented, white flowers, about 3 cm across, which are borne on short stalks above the rosette, open only on sunny days.

Flowering period	July to Oct.
Distribution	Widespread in northern and eastern suburbs and throughout the state; also SA.
Habitat	Open forest, heathy forest, woodland, heathland and grassland.
Notes	Often grows in extensive patches. When viewed from above, the showy flowers appear to squat on the rosette of leaves.
Similar species	None in the region.
Cultivation	Grown successfully by specialist growers.

Family Chenopodiaceae

Description

Scrambling or climbing shrub with long slender stems and broad, greyish leaves, to 3 cm x 1 cm. Tiny clusters of greenish flowers at the end of the branches are followed by showy clusters of succulent, red berries, ripening in autumn.

Flowering period	Aug. to Oct.
Distribution	Widespread in the Melbourne region and throughout the state; occurs in all states.
Habitat	Grows in a very wide range of habitats from open grassland to forests and rocky areas.
Notes	The slender stems thread through surrounding vegetation and the succulent fruit are eaten by birds. Previously known as *Rhagodia nutans*.
Similar species	*Rhagodia candolleana* has similar fruit but is much denser growing and with dark green leaves.
Cultivation	Requires well-drained soil in sun or filtered sun.

Family Cyperaceae

Description

Aquatic herb forming spreading patches, with woody rhizomes buried in mud, and bright green, erect, pithy, ribbed stems, to 1.5 m tall. These end in a cylindrical spike, to 6 cm long, which is narrower than the supporting stem. This spike changes from pale green to brown as the flowers mature.

Flowering period	Mainly Nov. to Feb., but sporadic at other times.
Distribution	Northern and eastern suburbs, and throughout the state; occurs in all states, NZ and PNG.
Habitat	Freshwater swamps, billabongs and slow-moving streams; also colonises dams.
Notes	Forms dense clumps in water to 5 m deep. It is an important refuge plant for fish and waterfowl. Aborigines used the stems for weaving.
Similar species	*E. acuta* is a smaller plant that has narrower stems but spikes thicker than their supporting stems.
Cultivation	Readily grown in ponds or dams.

Family Epacridaceae

Description

Slender shrub with long wiry stems which are often sparsely branched or unbranched, and clothed with small, dark green leaves, to 1.5 cm x 0.6 cm, which taper to a prickly tip. White or pink tubular flowers, each about 2 cm long, are borne in profusion along the length of the stems to produce an impressive floral display.

Flowering period	March to Nov.
Distribution	Widespread in the Melbourne region and throughout much of the state; also NSW, Tas, and SA.
Habitat	Open forest, woodland and heathland.
Notes	A distinctive popular plant with attractive nectar-rich flowers, which are commonly white but may also be in shades of pink or red. It is Victoria's floral emblem.
Similar species	None in the region.
Cultivation	Requires well-drained soil in filtered or dappled sun.

Family Epacridaceae

Description

Slender shrub with long, wiry, sparsely branched stems clothed with short, blunt leaves, to about 1 cm x 0.5 cm. White to cream, tubular or bell-shaped, fragrant flowers, to about 1 cm long, are carried in large numbers along the length of the stems.

Flowering period	June to Dec.
Distribution	*Rare* in the Melbourne region where known only from a couple of eastern suburbs, and disjunct in southern parts of the state; also Qld, NSW, Tas and SA.
Habitat	Wet or swampy areas in heathland and heathy forest.
Notes	A distinctive species which often grows in localised colonies.
Similar species	None in the region.
Cultivation	Readily grown in a container, but can be difficult to grow in the ground.

Family Myoporaceae

Description

Bushy, freely branching shrub, to 4 m tall, with light green to dark green, thick, leathery leaves, to 5 cm x 0.8 cm, which have a hooked apex. Small, nodding, cream, bell-shaped flowers, about 1 cm across, borne in the upper axils and on the end of branchlets, are followed by ovoid, yellow fruit about 0.6 cm long.

Flowering period	Mainly May to Oct., but also sporadic at other times.
Distribution	Known from a few western suburbs, but *now reduced to rarity* in the region; also in western parts of the state and in Qld, NSW, SA and WA.
Habitat	Grassland and box woodland, often on heavy soils.
Notes	The fruit are poisonous to stock. Previously known as *Myoporum deserti*.
Similar species	None in the region.
Cultivation	Requires well-drained soil in a sunny position.

Family Orchidaceae

Description

Small terrestrial orchid, flowering while the leaf is only partially
developed. When fully developed the single, dark green leaf, to
3.5 cm x 0.8 cm, is ovate and ground-hugging. The thin flower-
stem, to 25 cm tall, bears one to five flowers, about 2 cm across,
which have two prominent, white or pinkish sepals projecting
from the centre of the flower below the tightly recurved lip.

Flowering period	March and April.
Distribution	Northern and eastern suburbs, and throughout much of the state; also Qld, NSW, ACT, Tas and SA.
Habitat	Open forest, heathy forest, woodland and heathland.
Notes	This is usually among the first of the autumn orchids to flower.
Similar species	None in the region.
Cultivation	Grown successfully in containers by orchid specialists.

Family Apiaceae

Description

Perennial herb growing in spreading patches, each plant with a carrot-like rootstock and a tussock of bright green, prickly leaves, to 20 cm long. Small, pale blue flowers are carried in dense, globular heads on short, branched inflorescences, sometimes in a massed display.

Flowering period	Oct. to Feb.
Distribution	Widely, but disjunctly, distributed in the Melbourne region and in lowland areas of the state; also Qld, NSW, Tas, SA, WA and NZ.
Habitat	Low-lying seasonally wet depressions and swamps in open forest, woodland and grassland.
Notes	A distinctive easily recognised species with prickly leaves.
Similar species	*E. rostratum* has taller flower stems, bright blue flowers and deeply lobed leaves.
Cultivation	Readily grown in a container with the base immersed in water.

Family Myrtaceae

Description

Tree, to 40 m tall, with a dense, often widely spreading canopy and a straight or crooked trunk, covered with thick, brown, stringy bark. Dull, often hairy juvenile leaves, to 13 cm x 8 cm, are broadly ovate, whereas adult leaves, to 13 cm x 3 cm, are hairless, sickle-shaped, dark green and leathery. Axillary clusters of 7–15, club-shaped buds, open to showy white flowers, followed by capsules to 1.5 cm across, with slightly protruding valves.

Flowering period	Dec. to April.
Distribution	Eastern suburbs and widespread in the state; also SA.
Habitat	Moist open forest.
Notes	In open situations plants develop a rounded crown.
Similar species	*E. obliqua* has glossy, hairless juvenile leaves and the fruit valves are sunken.
Cultivation	Readily grown in a range of soil types and positions.

Family Myrtaceae

Description

Tree, to 50 m tall, with an open, spreading habit and a crooked trunk covered with smooth, greyish, pinkish, bluish or whitish bark, which sheds annually. Juvenile leaves, to 26 cm x 8 cm, are bluish-green, whereas the adult leaves, to 25 cm x 2 cm, are usually bright green to dark green. Axillary clusters of 7–11 buds, with sharply beaked points, open to white flowers, followed by capsules, to 1 cm across, with protruding valves.

Flowering period	Nov. to March.
Distribution	Widespread in the Melbourne region and throughout the state; also Qld, NSW, ACT, SA, WA and NT.
Habitat	Streamside vegetation and moist to wet areas in woodland and grassland.
Notes	A valuable tree which yields durable timber, excellent firewood and tasty honey.
Similar species	*E. blakelyi* is a smaller denser tree and the conical buds taper from near the centre to the apex.
Cultivation	Too large for a home garden but widely planted in regeneration projects.

Family Myrtaceae

Description

Tree, to 20 m tall, with a dense, often spreading canopy and a straight or crooked trunk, covered with thick, grey stringy bark. Juvenile leaves, to 11 cm x 6 cm, in opposite pairs, are stalkless, rounded and greyish-green, whereas the greyish-green mature leaves, to 20 cm x 2.5 cm, are distinctly stalked and alternate. Axillary clusters of 7–11, silvery or whitish, diamond-shaped buds, open to showy cream flowers, followed by flat-topped, silvery capsules, to 0.8 cm across.

Flowering period	March to Aug.
Distribution	Widespread in the eastern suburbs and endemic to the south-eastern part of the state.
Habitat	Moist open forest and woodland.
Notes	Flushes of new growth are an attractive silvery grey and flowers can appear in a beautiful profusion.
Similar species	None in the region.
Cultivation	An adaptable ornamental tree which grows readily in heavy but well-drained soil.

Family Myrtaceae

Description

Tree, to 35 m tall, with a moderately dense, rounded canopy and a straight trunk covered with thick, brown stringy bark. Hairy juvenile leaves, to 5 cm x 3 cm, have wavy margins and are shortly stalked, whereas the adult leaves, to 15 cm x 5 cm, are dark green and glossy. Axillary clusters of 7–11 diamond-shaped buds on a flattened stalk, open to white or cream flowers, followed by globose capsules to 1.2 cm across, with protruding valves.

Flowering period	Jan. to April.
Distribution	Northern and eastern suburbs, and widespread, especially in the east of the state; also NSW, ACT and SA.
Habitat	Drier types of open forest.
Notes	This tree is valued as a source of honey and its flowers have a strong honey scent. It is very sensitive to eucalypt dieback disease.
Similar species	*E. globoidea* has smaller leaves and fruit.
Cultivation	Easily grown but requires excellent drainage.

Family Myrtaceae

Description

Tree, to 25 m tall, with an open, sparse to dense canopy, and a straight trunk covered with fibrous, grey, yellow or brown, tessellated bark, the branches smooth and often whitish or yellowish. Juvenile leaves, to 11 cm x 5 cm, are greyish-green and oval, whereas the adult leaves, to 14 cm x 2.5 cm, range from green to slate-grey or bluish. Axillary clusters of five to seven, club-shaped buds, open to white flowers, followed by globose, flat-topped capsules, to 0.7 cm across.

Flowering period	Sept. to March.
Distribution	Widespread in the Melbourne region and over much of the state, especially in the east; also Qld, NSW and ACT.
Habitat	Drier types of open forest and woodland.
Notes	One of the finest sources of honey and also much valued for its shelter, timber and fuel wood.
Similar species	None in the region.
Cultivation	Readily grown in a range of situations and well-drained soil types.

107

Family Myrtaceae

Description

Tree, to 25 m tall, with an open, spreading canopy and a crooked or straight trunk, covered with light grey, tessellated bark. Oval, greyish-green juvenile leaves, to 11 cm x 5 cm, contrast with the dark green adult leaves, to 15 cm x 2 cm. Terminal clusters of seven to nine, diamond-shaped, yellowish buds, open to white flowers, followed by small, cup-shaped capsules, to 0.5 cm across.

Flowering period	Feb. to Aug.
Distribution	Western and northern suburbs, and widespread in the west and north of the state; also Qld, NSW and SA.
Habitat	Box eucalypt woodland.
Notes	This tree is valued for its timber and as a source of firewood and honey.
Similar species	*E. albens* has bluish-green leaves and buds.
Cultivation	Grows best in heavy soils.

Family Myrtaceae

Description

Tree, to 30 m tall, with a narrow to spreading, rather open canopy
and a straight trunk covered with rough bark at the base, and smooth,
grey or whitish bark further up, often shedding in strips. Juvenile
leaves, to 10 cm x 7 cm, are dull green, whereas the adult leaves, to
15 cm x 3 cm, are glossy green and often have wavy margins. Axillary
clusters of seven, diamond-shaped buds, open to white flowers,
followed by thick, flat-topped capsules, to 0.7 cm across.

Flowering period	March to June.
Distribution	Northern and eastern suburbs, and throughout much of the southern half of the state; also NSW, SA and Tas.
Habitat	Moist forests, streamside vegetation and swampy depressions.
Notes	This species is a food tree of koalas.
Similar species	*E. camphora* has broader leaves and smaller fruit with protruding valves.
Cultivation	An excellent tree for wet soils.

Family Myrtaceae

Description

Tree, to 25 m tall, with a dense, often rounded crown and straight or crooked trunk, covered with rough bark at the base, and grey or brown, tessellated bark higher up. Greyish-green juvenile leaves, to 6.5 cm x 8 cm, are rounded with a notched apex, whereas the adult leaves, to 9 cm x 3 cm, are slate grey or bluish-green. Terminal clusters of seven, club-shaped, yellowish buds, open to white flowers, followed by flat-topped, barrel-shaped, bluish capsules, to 0.4 cm across.

Flowering period	Sept. to Jan.
Distribution	Mainly northern and north-eastern suburbs, and northern parts of the state; also NSW.
Habitat	Drier types of open forest and box eucalypt woodland.
Notes	A highly ornamental tree valued as a source of honey and firewood.
Similar species	None in the region.
Cultivation	Requires excellent drainage in a sunny location.

Family Myrtaceae

Description

Tree, to 15 m tall, with a rounded to widely spreading, often dense crown, and a short crooked trunk covered with grey-brown fibrous bark, the upper branches being smooth. Dark green juvenile leaves, to 15 cm x 2.5 cm, are stalkless, whereas the adult leaves, to 18 cm x 2 cm, are stalked. Axillary groups of three ovoid buds, open to white flowers, followed by cup-shaped capsules, to 0.9 cm across, with protruding valves.

Flowering period	March to May.
Distribution	South-eastern suburbs and endemic in south-eastern parts of the state.
Habitat	Coastal and near-coastal forests and heathy forest, usually on sandy soils.
Notes	This species is often retained in parkland and around golf courses. It is valued as a source of firewood and honey and the leaves are eaten by koalas.
Similar species	*E. viminalis* is taller, with a long straight trunk and sheds bark from its trunk and large branches in long ribbons.
Cultivation	Excellent for coastal districts.

Family Myrtaceae

Description

Tree, to 30 m tall, with a rounded, dense crown, often with weeping branchlets, and a short trunk covered with grey, finely fissured, fibrous bark. Green, thin-textured juvenile leaves, to 18 cm x 3.5 cm, are stalkless and in opposite pairs, whereas the adult leaves, to 15 cm x 1.5 cm, are stalked. Axillary clusters of 7–16, club-shaped buds, open to white flowers, followed by cup-shaped capsules, to 0.6 cm across, with the valves enclosed.

Flowering period	Oct. to Jan.
Distribution	Widespread in the Melbourne region and in eastern parts of the state; also NSW and ACT.
Habitat	Moist areas in open forest and woodland.
Notes	This species is valued for its appealing shape and pleasantly aromatic leaves.
Similar species	*E. dives* has broader bluish-green juvenile leaves and broader adult leaves.
Cultivation	Adaptable to a range of soils and positions.

Family Myrtaceae

Description

Tree, to 25 m tall, with an elongated to spreading, rather open crown, and a short to tall, straight trunk, the base covered with persistent grey bark, the upper parts with white bark shedding in long, hanging ribbons. Stalkless juvenile leaves, to 6.5 cm x 2 cm, are in opposite pairs, whereas the dark green adult leaves, to 20 cm x 2 cm, are alternate and stalked. Axillary groups of three ovoid buds, open to white flowers, followed by cup-shaped fruit, to 0.9 cm across, with protruding valves.

Flowering period	Jan. to May.
Distribution	Widespread in the Melbourne region and throughout much of the state; also Qld, NSW, ACT, Tas. and SA.
Habitat	Moist forests particularly in valleys and near streams.
Notes	This species is valued as a source of timber, honey and food for koalas.
Similar species	*E. pryoriana* is shorter with a crooked trunk covered with fibrous bark.
Cultivation	Suitable for parkland and reclamation planting.

Family Santalaceae

Description

Shrub or small tree, to 8 m tall, and often forming colonies by root suckers. The plants have an upright to spreading, bushy habit and commonly take on a bronze to yellowish appearance. The branchlets, which may be erect or weeping, carry tiny, scale-like leaves and minute flowers, followed by a globular green fruit attached to a swollen, bright red, succulent stalk.

Flowering period	Mainly Aug. to Dec, but also sporadic at other times.
Distribution	Northern and eastern suburbs, and widely distributed throughout the state; also Qld, NSW, ACT, Tas, and SA.
Habitat	Open forest and woodland where it is often prominent on road cuttings.
Notes	This plant is parasitic on the roots of other plants. The red fleshy fruit stalk is edible.
Similar species	None in the region.
Cultivation	Virtually impossible to grow.

Family Frankeniaceae

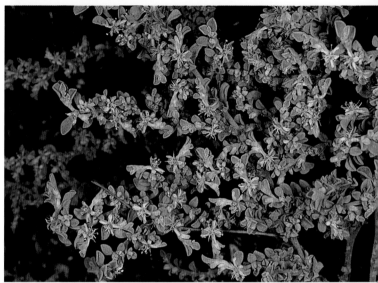

Description

Widely spreading prostrate shrub with a mat-like appearance and tangled stems covered with small, oblong, greyish-green leaves. Tiny pink flowers, about 0.3 cm across, are borne in clusters at the end of the branchlets, sometimes in profuse displays.

Flowering period	Sporadic all year.
Distribution	Eastern and western bayside suburbs.
Habitat	Coastal dunes and headlands and coastal saltmarshes.
Notes	An important coastal plant tolerating buffeting salt-laden winds and saline soils.
Similar species	None in the region.
Cultivation	A useful plant for coastal districts.

Family Cyperaceae

Description

Perennial herb forming individual tufts or clumps, and spreading by underground rhizomes. The dark green, rough-textured leaves, to 2 m tall, have sharp cutting edges. Flower-stems, about as long as the leaves, carry a terminal panicle with dark brown to black pendulous segments and similar-coloured small flowers.

Flowering period	Sporadic all year.
Distribution	Northern and eastern suburbs, and throughout much of the state; also Qld, NSW, Tas and SA.
Habitat	Drier types of open forest and woodland.
Notes	This species dominates the ground layer of some forests. The leaves have been used for roof thatching.
Similar species	Other species of *Gahnia* in the region form large tussocks.
Cultivation	Resents disturbance and difficult to establish. Best raised from seed.

Family Orchidaceae   Cinnamon Bells, Potato Orchid

Description

Leafless terrestrial orchid with a thick subterranean rhizome and brown flower-stems, to 50 cm tall, which nod in bud. The apex of each stem is a loose open raceme of 4–20 pendulous, tubular flowers, each about 2 cm long. These are cinnamon brown with white tips and have a pleasant spicy scent. The lip is enclosed within the flower.

Flowering period	Oct. to Dec.
Distribution	Eastern suburbs, Dandenong Ranges, and scattered throughout the state; also Qld, NSW, ACT, Tas and SA.
Habitat	Moist situations in open forest, heathy forest and woodland.
Notes	Flowering is promoted by summer fires.
Similar species	*G. procera* is much more robust and the flower stems do not nod in bud.
Cultivation	Impossible to grow.

Family Orchidaceae

Description

Slender terrestrial orchid, to 25 cm tall, with a very thin leaf and flower-stem, the latter ending in a dense spike of 5–40 crowded, upside-down flowers. Each flower, which is only about 0.3 cm across, is nodding, dark purplish-brown and with pointed segments. All parts are hairless and the lip is sharply pointed.

Flowering period	Feb. to April.
Distribution	Northern and eastern suburbs, Dandenong Ranges, and scattered throughout the state; also NSW, Tas and SA.
Habitat	Open forest, woodland, heathy forest and heathland.
Notes	An inconspicuous orchid which is readily overlooked.
Similar species	*G. aff. rufum* has blunt segments.
Cultivation	There has been little success in cultivating this plant.

Family Orchidaceae

Description

Slender terrestrial orchid, to 30 cm tall, with a very thin leaf and flower-stem, the latter ending in a dense spike of 5–15 crowded, upside-down flowers. Each flower, which is about 0.8 cm across, is semi-nodding, dark purplish, and with prominent coarse hairs on the lip and some other segments.

Flowering period	Feb. to March.
Distribution	Eastern suburbs, Dandenong Ranges, and scattered throughout the state; also Tas.
Habitat	Open forest, heathy forest and heathland.
Notes	The lip is delicately hinged and trembles in the slightest breeze.
Similar species	*G. archeri* has smaller flowers with very short hairs on the lip.
Cultivation	There has been little success in cultivating this plant.

Family Geraniaceae

Description

Prostrate, herb-like shrub with slender, spreading branches which often thread through surrounding shrubs. The rounded, hairy, dull green leaves, to 7 cm x 5 cm, borne on long slender stalks, are incised almost to the base into five to seven segments, which are themselves toothed. Pink or purplish flowers, 1–1.5 cm across, are borne singly towards the end of the stems.

Flowering period	Sept. to March.
Distribution	Eastern suburbs and throughout much of the state; also NSW, ACT, Tas, SA and NZ.
Habitat	Moist areas in grassland, open forest and woodland.
Notes	This species is variable in flower colour, with some plants having nearly white flowers.
Similar species	*G. solanderi* bears its flowers in pairs.
Cultivation	Easily grown in a moist sheltered position.

Family Gleicheniaceae

Description

Scrambling or climbing fern which forms dense tangled thickets, to 4 m tall. The much-branched, wiry fronds thread through surrounding shrubbery, supporting spreading segments clothed with tiny, bead-like, flat leaflets.

Distribution	Eastern suburbs, Dandenong Ranges, and in southern parts of the state; also NSW, Tas, SA, NZ, NCal and Asia.
Habitat	Boggy areas and soaks in open forest and woodland; sometimes colonising disturbed sites.
Notes	An interesting fern with a highly specialised growth habit. Plants in exposed situations appear yellowish, whereas those in a shady position are bright green.
Similar species	*G. dicarpa* has strongly concave or pouched leaflets.
Cultivation	Resents disturbance and is very difficult to establish.

Family Orchidaceae

Description

Small terrestrial orchid with a broad hairy leaf, to 6 cm x 2 cm, often flat on the ground, and one or two conspicuous, mauve or purple flowers, 3.5–4.5 cm across, on a slender, hairy stalk to 30 cm tall. The flowers have a recurved rear sepal and four other spreading segments. The lip is white at the base with a mauve or purple tip.

Flowering period	Oct. to Nov.
Distribution	Northern and eastern suburbs, Dandenong Ranges, and throughout much of the state; also Qld, NSW, ACT, Tas and SA.
Habitat	Open forest, woodland, heathy forest and heathland.
Notes	A familiar species which is often locally common and produces colourful floral displays in good seasons.
Similar species	None in the region.
Cultivation	There has been little success in cultivating this plant.

Family Fabaceae

Description

Slender climber with thin twining stems, to 2 m tall, arising from a carrot-like rootstock, and bearing hairy, dull green, trifoliolate leaves with narrow leaflets, to 4 cm long, on long, slender stalks. Small, bluish-mauve, pea-shaped flowers, to 0.8 cm across, carried in slender axillary racemes, are followed by small pods.

Flowering period	Oct. to Feb.
Distribution	Northern and eastern suburbs, and throughout the state; also Qld, NSW, ACT, Tas, SA and WA.
Notes	This species becomes noticeable only at flowering time.
Similar species	*G. latrobeana* is more shrubby with trailing non-climbing stems.
Cultivation	An excellent small climber for mingling with shrubs.

Family Fabaceae

Description

Twiggy spreading shrub, to 1 m tall, with numerous slender branches clothed with bluish-green, trifoliolate leaves, each with narrow pointed leaflets, to 2.5 cm long. Showy, bright yellow, pea-shaped flowers, about 2 cm across, borne in small groups from the end of each branchlet, are followed by swollen pods, to 1.5 cm long.

Flowering period	Sept. to Feb.
Distribution	Eastern suburbs, Dandenong Ranges, and throughout much of the state; also Qld, NSW, ACT and SA.
Habitat	Open forest, heathy forest and heathland.
Notes	An extremely showy species with large, colourful flowers.
Similar species	None in the region.
Cultivation	Grows well in containers. Requires excellent drainage and filtered sun or partial sun.

Family Goodeniaceae

Description

Shrub with a bushy to open growth habit, and light green to yellowish-green, ovate leaves, to 10 cm x 6 cm, with coarsely toothed margins and prominent venation. Bright yellow flowers, about 1 cm across, are carried in clusters of one to six from the upper leaf axils.

Flowering period	Mainly Aug. to Feb., but also sporadic at other times.
Distribution	Widespread in the Melbourne region and throughout the state; also Qld, NSW, ACT, Tas, SA and NT.
Habitat	Coastal scrub, heathland, open forest and woodland.
Notes	A fast-growing species which often colonises sites of disturbance.
Similar species	None in the region.
Cultivation	Readily grown in a range of soils and positions. Needs regular tip pruning.

125

Family Fabaceae

Description

Open shrub, to about 5 m tall, often growing in colonies, with pinkish branchlets bearing trifoliolate leaves which have bluish-green, ovate leaflets, to 2.5 cm x 1.5 cm. Pea-shaped flowers, about 1 cm across, which are yellow with reddish markings, are borne in showy racemes from near the end of the branches.

Flowering period	Sept. to Dec.
Distribution	Mostly in the eastern suburbs and Dandenong Ranges, but with occurrences in the west and north, and throughout much of the state; also in Qld, NSW, ACT, Tas, SA and WA.
Habitat	Moist areas of open forest and woodland.
Notes	A very fast-growing species which can spread through suckers. Seed germinates profusely after fire.
Similar species	None in the region.
Cultivation	Easily grown but benefits from regular pruning.

Family Scrophulariaceae

Description

Perennial suckering herb with erect to sprawling, soft, fleshy stems and opposite pairs of light green, ovate leaves, to 4.5 cm x 2.5 cm, with toothed margins. Pink to purplish, tubular flowers, about 1 cm across, are borne in the leaf axils.

Flowering period	Oct. to May.
Distribution	Widespread in the Melbourne region and throughout the state; also NSW, ACT, Tas, SA, WA and NZ.
Habitat	Swamps, bogs and stream margins.
Notes	The stems form roots from the nodes as well as producing basal suckers.
Similar species	*G. pubescens* has smaller densely hairy leaves.
Cultivation	Excellent for ponds or waterlogged areas in the garden.

Family Proteaceae

Description

Stiff shrub, to 3 m tall, with an open to dense habit, hairy, yellowish new growth, and dark green, crowded, cylindrical leaves, to 5 cm x 0.2 cm, with a black, pointed tip. Deep yellow, somewhat smelly flowers are carried in dense axillary clusters, followed by ovoid to globular, short-beaked, woody fruit, to 3.5 cm x 3 cm.

Flowering period	April to Aug.
Distribution	Northern and eastern suburbs, and in southern parts of the state; also Tas and SA.
Habitat	Swampy areas of heathland, heathy forest, open forest and woodland.
Notes	A vigorous species which grows rapidly after fire.
Similar species	*H. sericea* has much larger white or pink flowers.
Cultivation	Grows readily and tolerates moist to wet soils.

Family Proteaceae

Description

Stiff shrub, to 5 m tall, with an open to dense habit, light green, hairy new growth and dark green, cylindrical leaves, to 8 cm x 0.2 cm, with a sharp, black tip. Dense axillary clusters of white, hairy, fragrant flowers, are followed by ovoid, dark brown, rough, woody fruit, to 3 cm x 2.5 cm, with a broad beak.

Flowering period	May to Oct.
Distribution	Northern and eastern suburbs, and widespread throughout much of the state; also NSW and Tas.
Habitat	Moist areas of open forest and woodland.
Notes	This species germinates profusely after fire and grows rapidly. Pink-flowered variants occur in some areas.
Similar species	*H. nodosa* has smaller clusters of yellow flowers.
Cultivation	Grows readily and may become weedy. An excellent refuge plant for small birds.

Family Proteaceae

Description

Stiff, often narrow, extremely dense shrub, to 5 m tall, with densely hairy new growth and dark green, extremely stiff, narrow leaves, to 18 cm x 0.4 cm, with a sharp, rigid point. Dense axillary clusters of white to cream, hairy flowers are followed by small, ovoid, short-beaked, woody fruit, to 2.5 cm x 1.5 cm.

Flowering period	July to Nov.
Distribution	Northern and eastern suburbs, and widespread in southern areas of the state; also NSW and Tas.
Habitat	Heathland, heathy forest and open forest.
Notes	This species, which often grows in thickets, regenerates strongly from seed.
Similar species	None in the region.
Cultivation	Grows readily in well-drained to moist soils.

Hardenbergia violacea Sarsaparilla, Purple Coral Pea

Family Fabaceae

Description

Vigorous bushy climber with a thick, woody rootstock and long, thin, twining stems clothed with broad, leathery, dark green leaves, to 12 cm x 5 cm, with prominent veins. Masses of purple, pea-shaped flowers, about 1 cm across, with a prominent, green central eye, are borne in sprays on long racemes, followed by flat, leathery, dark brown to blackish pods.

Flowering period	July to Nov.
Distribution	Widespread in the Melbourne region and throughout the state; also in Qld, NSW, ACT, Tas and SA.
Habitat	Open forest and woodland, often on stony or gravelly slopes and ridges.
Notes	A familiar plant which produces colourful floral displays.
Similar species	None in the region.
Cultivation	Requires excellent drainage in sun or partial sun. Many selections are grown by nurseries.

Family Asteraceae

Description

Perennial herb spreading by vigorous rhizomes to form large, often dense patches, with grey woolly new growth. The erect, hairy stems, to 50 cm tall, which branch above the base and bear stem-clasping, hairy, greyish-green leaves, to 8 cm x 0.8 cm, end in a single, bright yellow, button-like flower-head, 1.5–2 cm across.

Flowering period	Sept. to April.
Distribution	Scattered in western, northern and eastern suburbs, and widely distributed in the state; also Qld, NSW, ACT and SA.
Habitat	Moist to wet areas of grassland and woodland.
Notes	In wet seasons the plants can have an extended flowering period with flowers developing as new growth matures.
Similar species	*H. scorpioides,* which is generally less vigorous, has stems which branch at soil level and non-stem-clasping leaves.
Cultivation	Readily grown in a sunny position in moist well-drained soil.

Family Asteraceae

Description

Perennial herb spreading by rhizomes to form a compact clump, with woolly new growth. The erect to sprawling, hairy stems, to 30 cm long, which branch at ground level and bear non-stem-clasping, grey, velvety leaves, to 10 cm x 0.8 cm, end in a single, flat, pale yellow flower-head, 2–3 cm across.

Flowering period	Sept. to Dec.
Distribution	Widespread in the Melbourne region and throughout much of the state; also Qld, NSW, ACT, Tas and SA.
Habitat	Grassy and shrubby forests, woodland and heathland.
Notes	The basal leaves are largest and the leaf size diminishes up the stem. Plants may die back after flowering and reshoot in autumn.
Similar species	*H. rutidolepis,* which is more vigorous, has stems branching above ground level and stem-clasping leaves.
Cultivation	Grows well as a container plant or among rocks.

Family Dilleniaceae

Description

Trailing or climbing shrub, with long, wiry, reddish stems and small, dark green leaves, to 0.8 cm x 0.4 cm, which are rough to the touch. Masses of bright yellow flowers, each about 1 cm across, on slender stalks to 1.5 cm long, are borne in the upper axils and on the end of short branchlets.

Flowering period	Mainly Aug. to Oct., but also sporadic at other times.
Distribution	Rare in the region where known only from a few eastern suburbs and the Dandenong Ranges, but widespread in southern parts of the state; also Qld, NSW, Tas and SA.
Habitat	Moist areas of open forest, often in valleys.
Notes	The long stems of this species can thread through surrounding vegetation to a height of about 2 m. *H. astrotricha* and *H. billardieri* are synonyms.
Similar species	None in the region.
Cultivation	Easily grown in a moist sheltered position.

Family Dilleniaceae

Description

Shrub, to about 0.5 m tall, with an erect or spreading habit, twiggy branches, and broad, blunt, greyish-green leaves, to 2 cm x 1 cm. Bright golden yellow flowers, 2–3 cm across, with prominently notched petals are carried in profusion in the upper axils.

Flowering period	Mainly Aug. to Feb., but also sporadic at other times.
Distribution	Eastern suburbs, Dandenong Ranges, and widespread throughout the state; also Qld, NSW, ACT and Tas.
Habitat	Open forest and woodland.
Notes	This species, although variable in growth habit, always produces colourful floral displays.
Similar species	*H. stricta* has much narrower leaves.
Cultivation	Requires excellent drainage in a sunny location.

Family Dennstaedtiaceae

Description

Coarse ground fern forming widely spreading clumps, with vigorous slender rhizomes and pale green to bluish-green fronds, to 2 m x 1 m. Each frond has a stout, brown stalk and broadly triangular blade divided three or four times. The segments are in opposite pairs and the lowermost pairs of each group form an outline resembling bats's wings. The spores are borne in narrow marginal bands.

Distribution	Eastern suburbs, Dandenong Ranges, and widespread in the state; also Qld, NSW, ACT, Tas, SA and NT and in many other countries.
Habitat	Moist valleys, sheltered slopes and fern gullies in tall forest; also colonises disturbed sites.
Notes	A distinctive fern which in suitable sites develops into vigorous spreading patches.
Similar species	None in the region.
Cultivation	Readily grown in moisture-retentive soil in a sheltered position.

Family Fabaceae

Description

Shrub with a woody rootstock and a few slender, erect to sprawling branches, to about 50 cm long. These have narrow, dull green leaves, which are often reflexed backwards, and clusters of mauve, pea-shaped flowers, each about 1 cm across, with a prominent cream to yellow central patch. Short, plump, distinctly hairy pods are black when ripe.

Flowering period	Aug. to Oct.
Distribution	Eastern suburbs, Dandenong Ranges, and throughout much of the state; also Qld, NSW, Tas and SA.
Habitat	Open forest, heathy forest and heathland.
Notes	A familiar species which regenerates freely after fires.
Similar species	None in the region.
Cultivation	Requires excellent drainage in partial sun or filtered sun.

Family Hymenophyllaceae

Description

Small epiphytic fern, usually found growing on tree fern trunks, with threadlike branching rhizomes and drooping, pale green, membranous fronds, usually 5–20 cm long (but occasionally to 30 cm). The fronds, which are variable in shape, have numerous narrow lobes with broad spore-bearing segments at the apex.

Distribution	In the region confined to the Dandenong Ranges, and in southern parts of the state; also NSW, Tas, NZ and Polynesia.
Habitat	Decaying logs, butts of large trees and fibrous tree fern trunks in shaded fern gullies.
Notes	A delicate fern with fronds only one cell thick and unable to withstand desiccation.
Similar species	*H. cupressiforme* and *H. australe* both have much smaller fronds.
Cultivation	Suitable only for a terrarium.

Family Dennstaedtiaceae

Description

Coarse ground fern forming widely spreading clumps, with vigorous slender rhizomes and pale green to yellowish-green, lacy fronds, to 1.5 m x 70 cm. Young parts, especially developing fronds, are covered with sticky hairs. Each frond has a stout, brown hairy stalk and triangular, softly hairy blade which is divided two or three times. The spores are borne in small circular clusters.

Distribution	Eastern suburbs, Dandenong Ranges, and in south-eastern parts of the state; also Qld, NSW, ACT, NCal and PNG.
Habitat	Moist to swampy areas in tall forests and near streams.
Notes	In suitable conditions this species can form extensive spreading patches.
Similar species	*H. muelleri* has stiff dark green fronds; *H. rugosula* has warty frond stalks and pale green fronds.
Cultivation	Readily grown in moist soil.

Family Hypoxidaceae   Golden Weather Grass

Description

Small, perennial, bulbous, lily-like herb with a sparse tuft of a few narrow, dark green, grass-like leaves, to 25 cm long, and bright yellow, starry flowers, 2.5–3 cm across, on slender stems longer than the leaves. The flowering stems have a single bract near each flower. All parts of the plant have wispy whitish hairs.

Flowering period	Aug. to Feb.
Distribution	Northern and eastern suburbs, and throughout much of the state; also NSW, ACT and Tas.
Habitat	Moist grassy areas in open forest and woodland.
Notes	The showy flowers last more than one day, closing in late afternoon and remaining closed in cool or rainy weather.
Similar species	*H. glabella* and *H. vaginata* which are both hairless.
Cultivation	Readily grown in a container or among rocks in the garden.

Family Fabaceae

Description

Sparse shrub with an erect to spreading habit and long, slender branches clothed with soft, bluish-green, pinnate leaves, to 10 cm long. Long racemes, to 14 cm long, of conspicuous, pink to mauve, pea-shaped flowers, each about 1 cm across, arise in the upper axils and are followed by slender, cylindrical, angular, brown pods, to 4 cm x 0.3 cm.

Flowering period	Sept. to Dec.
Distribution	Widespread in the Melbourne region and in much of the state; occurs in all states.
Habitat	Moist areas of open forest, especially in valleys near streams.
Notes	A fast-growing species which regenerates strongly after fire.
Similar species	None in the region.
Cultivation	Grows readily in a moist, sheltered position. Best with regular pruning.

Family Cyperaceae

Description

Perennial herb forming crowded tussocks of erect, yellowish-green to dark green, wiry stems, to 1.5 m tall, each with a dense, globular, brown flower-head, about 2 cm across, borne just below the apex.

Flowering period	Sporadic all year.
Distribution	Widespread in the Melbourne region and throughout much of the state; occurs in all states and many overseas countries.
Habitat	Coastal scrubs on dunes and headlands, heathy forest, heathland and saltmarshes.
Notes	An interesting species which is becoming widely planted as a component of revegetation programs in eastern bayside suburbs. Previously well known as *Scirpus nodosus*.
Similar species	None in the region.
Cultivation	Can be grown in a wide range of soils and positions.

Family Proteaceae

Description

Small shrub, to 50 cm tall, with a woody rootstock and a compact mound of stiff, rigid, prickly, light green leaves, to 8 cm x 5 cm, which are divided into numerous narrow segments. Small, bright yellow flowers are carried in prominent heads, to 3 cm across, among the upper leaves.

Flowering period	Sept. to Nov.
Distribution	Occurs in a few south-eastern suburbs and *now rare in the region*; also mainly in southern parts of the state, NSW, Tas and SA.
Habitat	Heathland and heathy forest.
Notes	The range of this species around Melbourne has been greatly reduced by urbanisation.
Similar species	None in the region.
Cultivation	Requires excellent drainage and can be difficult to maintain.

Family Juncaceae

Description

Perennial rush which forms a dense tussock, to about 1 m tall, with slender green stems topped with tiny brown flowers arranged in loose to dense terminal clusters.

Flowering period	Nov. to March.
Distribution	Northern and eastern suburbs, and mainly in eastern areas of the state; also NSW and Tas.
Habitat	Moist to wet depressions and swamps in open forest and woodland.
Notes	A weedy species which can colonise drainage ditches and boggy sites.
Similar species	One of a group of similar-looking plants which can be very difficult to identify.
Cultivation	Readily grown in boggy soils.

Family Fabaceae

Description

Prostrate shrub with a woody rootstock and long, slender, trailing stems clothed with distinctive, greyish-green, trifoliolate leaves, which have hairy, crinkly leaflets, to 8 cm x 5 cm. Colourful, bright red to scarlet, pea-shaped flowers, to 2 cm across, which have a prominent yellowish central blotch, are followed by dark brown, leathery pods, about 7 cm long.

Flowering period	April to Dec.
Distribution	Widespread in the Melbourne region and throughout much of the state; also NSW, ACT, Tas, SA and WA.
Habitat	Grassland, open forest, woodland, heathy forest and heathland.
Notes	A familiar species which can produce colourful floral displays. Regeneration is vigorous after fire.
Similar species	None in the region.
Cultivation	Requires well-drained soils in a sunny or semi-shady location.

Family Myrtaceae

Description

Vigorous shrub or tree, to 8 m tall, which often grows in thickets.
Plants have a dense to open habit, erect to weeping branches, and
green to reddish leaves, to 2.5 cm x 0.4 cm. White to cream,
scented flowers, each about 1.2 cm across, are borne in masses
in long, leafy sprays towards the end of the branches.

Flowering period	Nov. to Feb.
Distribution	Northern and eastern suburbs, and throughout eastern parts of the state; also Qld, NSW and ACT.
Habitat	Open forest, often on clay slopes and flats near streams.
Notes	A variable species which is attractive when in flower. It is a vigorous coloniser of disturbed ground. Previously known as *Leptospermum phylicoides.*
Similar species	None in the region.
Cultivation	Readily grown in a range of soils and positions.

Family Dryopteridaceae

Description

Ground fern growing in neat discrete clumps, with a short creeping or erect rootstock, and a crown of arching, dark green, glossy fronds, to 80 cm x 30 cm. Each frond has a very slender, brown stalk and a triangular, blade, divided two or three times, which is somewhat rough to the touch. The spores are borne in small circular clusters.

Distribution	Dandenong Ranges and southern parts of the state; also Qld, NSW, ACT, Tas and SA.
Habitat	Moist valleys, sheltered slopes and fern gullies in tall forest.
Notes	This species usually grows in the ground but may also establish on decaying logs and tree fern trunks. Previously known as *L. shepherdii.*
Similar species	None in the region.
Cultivation	Grows readily in a moist sheltered position.

147

Family Malvaceae

Description

Shrubby herb, to 2.5 m tall, with soft to pithy, hairy branches and large, rounded, dark green, five to seven-lobed leaves, to 12 cm across, on long slender stalks. White, pink or lilac, prominently veined, hollyhock-like flowers, 4–6 cm across, are carried in axillary clusters and followed by flat, rounded, segmented fruit.

Flowering period	Mainly July to Sept., but also sporadic at other times.
Distribution	Scattered in western and northern suburbs close to Port Phillip Bay, and widespread in the state; occurs in all states.
Habitat	Grassland, salt marshes, coastal vegetation and wasteland.
Notes	This species, which may be fast growing and short-lived, sometimes colonises disturbed sites.
Similar species	None in the region.
Cultivation	Readily grown in a range of soil types in a sunny position.

Family Cyperaceae

Description

Perennial sedge forming dense spreading clumps, with tufts of erect, yellowish-green to green leaves which are flat on one side and curved on the other. The flower stems, to 1 m tall, are similar but end in a crowded panicle of tiny, brown flowers.

Flowering period	Sporadic all year.
Distribution	Eastern and south-eastern bayside suburbs and coastal districts of the state; also Qld, NSW and Tas.
Habitat	Coastal dunes and coastal heath.
Notes	An important species for stabilising coastal dunes. It often grows in dense thickets.
Similar species	*L. gladiatum* has the stems curved on both surfaces.
Cultivation	Requires excellent drainage. Useful in coastal districts.

Family Cyperaceae

Description

Perennial sedge forming large tussocks, with flat, dark green, shiny leaves, to 1.5 m x 1 cm, and flat stems, to 2 m tall, ending in a large panicle of brown flowers. The leaves have a reddish base.

Flowering period	Sporadic all year.
Distribution	Hilly eastern suburbs, Dandenong Ranges, and in southern parts of the state; also NSW.
Habitat	Moist slopes and gullies in tall forest.
Notes	The leaves were used by the Aborigines for weaving.
Similar species	*L. laterale var. laterale* grows to 1 m tall, has leaves to 0.6 cm wide and smaller panicles.
Cultivation	Requires a moist sheltered position.

Family Orchidaceae

Description

Small terrestrial orchid with one or two ovate, basal leaves, to 3 cm
x 2 cm, which are greyish-green or yellowish-green with prominent
red stripes. A thin, wiry flower-stem, to 25 cm tall, bears one to
three brownish-green flowers, about 1.5 cm across, which have two
petals held erect like the ears of a hare. The delicately hinged lip has
reddish markings and deeply fringed margins.

Flowering period	April to June.
Distribution	Eastern bayside suburbs, but *now reduced to great rarity;* also occurs in western parts of the state and SA and WA.
Habitat	Heathland and heathy forest in sandy soils.
Notes	This species often grows in large colonies with a very low proportion of flowering plants. Flowering can be stimulated by fires.
Similar species	None in the region.
Cultivation	There has been little success in cultivating this plant.

Family Orchidaceae

Description

Small terrestrial orchid with a single bright green, ovate, basal leaf, to 9 cm x 3 cm, and a thin, wiry flower-stem, to 25 cm tall, which carries one to three, pink to red and white flowers. These are of an unusual shape with two broad white sepals projecting forwards and two long, red, ear-like petals erect behind the flower. The lip has conspicuous red bars and yellow calli.

Flowering period	Sept. to Nov.
Distribution	Northern and eastern suburbs, and scattered throughout the state; also NSW, Tas, SA and WA.
Habitat	Open forest, heathy forest, woodland and heathland.
Notes	This species grows in dense colonies. Flowering is strongly promoted by summer fires.
Similar species	None in the region.
Cultivation	There has been little success in cultivating this plant.

152

Family Asteraceae

Description

Perennial herb forming bushy clumps, to about 30 cm tall, with wiry stems clothed with narrow, dark green leaves to 2 cm long, which are woolly-white beneath. Bright yellow, button-like flower-heads, 1–1.5 cm across, are carried singly on the end of long, slender, scaly stems.

Flowering period	Mainly Sept. to Jan., but also sporadic at other times.
Distribution	Widespread in the Melbourne region and throughout the state; also Qld, NSW, Tas and SA.
Habitat	Grassland, open forest and grassy woodland.
Notes	This species often has a compact habit and the well-displayed flower-heads are most noticeable.
Similar species	*L. tenuifolius* has longer leaves with the margins curved under.
Cultivation	Requires excellent drainage in a sunny position.

Family Myrtaceae

Description

Robust shrub, to 6 m tall, with a dense, bushy habit, silver-haired new growth and twiggy stems clothed with leathery, oblong leaves, to 3 cm x 0.8 cm, which are dark green on the upper surface and silky-haired beneath. Prominent white flowers, about 2 cm across, which are well displayed at the end of small branchlets, are followed by woody capsules, about 1 cm across.

Flowering period	Oct. to Jan.
Distribution	Northern and eastern suburbs, but scattered, and mainly in eastern parts of the state; also NSW.
Habitat	Moist forests, often near streams.
Notes	A robust species which has attractive new growth and prominent floral displays.
Similar species	*L. lanigerum* has smaller leaves and flowers about 1.5 cm across.
Cultivation	Readily grown in moist soil.

Family Myrtaceae

Description

Shrub or tree, to 8 m tall, with thick, greyish-brown bark and a dense, rounded to spreading canopy. Twiggy branchlets are clothed with thick, leathery, greyish-green leaves, to 3 cm long. White flowers, about 2 cm across, which are well-displayed on the end of very short side shoots, are followed by grey, woody capsules, about 0.8 cm across.

Flowering period	Mainly Aug. to Oct., but also sporadic at other times.
Distribution	Mainly eastern and south-eastern bayside suburbs, and in many areas of the state; also NSW, Tas and SA.
Habitat	In the region confined to coastal scrub, coastal headlands and heathy forest.
Notes	A familiar species which often dominates coastal vegetation. Old plants develop a characteristic gnarled, twisted trunk.
Similar species	None in the region.
Cultivation	Excellent for combating coastal exposure, but also adaptable to many other situations.

Family Myrtaceae

Description

Robust shrub, to 6 m tall, with a dense bushy habit, densely hairy new growth, and twiggy stems clothed with leathery, greyish-green, hairy leaves, to 1.5 cm x 0.4 cm. Prominent white flowers, about 1.5 cm across, well-displayed at the end of small branchlets, are followed by hairy, woody capsules, about 0.8 cm across.

Flowering period	Sept. to Jan.
Distribution	Widespread in the Melbourne region and throughout southern parts of the state; also NSW, Tas and SA.
Habitat	Streamside vegetation and moist to swampy depressions in open forest and heathy forest.
Notes	This species sometimes grows in dense thickets. The wood was used by the Aborigines to make spears.
Similar species	*L. grandifolium* has larger dark green leaves and flowers about 2 cm across.
Cultivation	Readily grown in a range of positions in moist soil.

Family Asteraceae

Description

Perennial herb forming sparse to dense clumps, with silvery new growth, short, brittle, hairy stems and crowded, narrow, grey, hairy leaves, to 10 cm x 0.5 cm. Bright yellow to orange, papery flower-heads, each 2 x 3 cm across, are borne singly on stalks, to 30 cm tall.

Flowering period	Nov. to March.
Distribution	Northern suburbs, but *now scattered and rare,* and in central and eastern parts of the state; also Qld and NSW.
Habitat	Usually among rocks or on gravelly slopes in open forest.
Notes	A showy species which can produce massed displays. Previously known as *Helipterum albicans.*
Similar species	None in the region.
Cultivation	Requires excellent drainage and free air movement in a sunny location. Can be short-lived.

157

Family Asteraceae

Description

Densely woolly shrub, which often grows in circular mounds, consisting mainly of slender, tangled branches covered by tiny, silvery-grey, scale-like leaves. New growth is whitish. Tiny, pale yellow flowers in globular heads, about 1 cm across, are borne on the end of the branchlets.

Flowering period	Sept. to Dec.
Distribution	Bayside suburbs and coastal areas of the state; also NSW, Tas, SA and WA.
Habitat	Coastal sand dunes, cliffs and headlands.
Notes	This species withstands dryness and buffeting by strong salt-laden winds. It also frequently grows in calcareous soils.
Similar species	None in the region.
Cultivation	Excellent for coastal gardens in a sunny location in well-drained soil.

Family Epacridaceae

Description

Erect shrub, to 1.5 m tall, with long, slender, wiry branches bearing narrow, prickly, dark green leaves, to 1 cm x 0.3 cm, with the margins curved under. Masses of tiny pink buds open to small white or pinkish cup-shaped flowers, which have small spreading hairy lobes.

Flowering period	July to Nov.
Distribution	Northern and eastern suburbs, especially near the coast, and throughout much of the state; also Qld, NSW, Tas and SA.
Habitat	Open forest, heathy forest and heathland.
Notes	This attractive species sometimes grows in local colonies.
Similar species	None in the region.
Cultivation	Requires excellent drainage in filtered sun or semi-shade.

Family Epacridaceae

Description

Bushy shrub, to 4 m tall, with dark grey, corrugated bark and pale green new growth. Twiggy stems are densely clothed in stiff, dark green leaves, to 2.5 cm x 0.6 cm, which are paler on the underside. Small, white, cup-shaped flowers with spreading hairy lobes, borne in dense masses, are followed by small, white, fleshy fruit.

Flowering period	July to Nov.
Distribution	Eastern bayside suburbs and coastal areas of the state; also Qld, NSW, Tas, SA, WA and NZ.
Habitat	Coastal scrub on dunes and headlands.
Notes	The flowers, which are often lightly tinged with pink, lack any scent. The fruit are edible but insipid.
Similar species	*L. australis* has angular branches, longer leaves and fragrant flowers.
Cultivation	Requires excellent drainage. Best grown in coastal districts.

Family Epacridaceae

Description

Erect shrub, to about 50 cm tall, with weakly erect or sprawling, twiggy branches clothed with stiff, narrow, dark green, hairy-margined leaves, to 2 cm x 0.3 cm. Small, fragrant, white, cup-shaped flowers, with spreading hairy lobes, are borne in conspicuous, densely-flowered spikes, to 2 cm long.

Flowering period	July to Dec.
Distribution	Eastern and south-eastern suburbs, and throughout much of the state; also Qld, NSW, ACT, Tas and SA.
Habitat	Open forest, heathy forest and heathland.
Notes	An attractive small species which is often locally common. At flowering time the whole plant can be covered in the small white flowers.
Similar species	None in the region.
Cultivation	Requires excellent drainage in filtered sun or semi-shade.

Family Lindsaeaceae

Description

Compact ground fern growing in small colonies, with distinctly different sterile and fertile fronds. The sterile fronds, which tend to be shorter and prostrate or spreading, have small, but well-developed, wedge-shaped leaflets with lobed margins. The fertile fronds, to 25 cm long, are held stiffly erect and have smaller, downcurved leaflets which appear to be arranged in a loose spiral.

Distribution	Widespread in the Melbourne region and throughout southern areas of the state; also Qld, NSW, Tas, SA, WA, NZ and NCal.
Habitat	Heathland, heathy forest, open forest and woodland, usually among small shrubs and herbs.
Notes	An interesting small fern which regenerates vigorously after fires, when it produces numerous fertile fronds.
Similar species	None in the region.
Cultivation	Resents disturbance and cannot be cultivated.

Family Epacridaceae

Description

Small shrub, to about 1 m tall, with an erect, rigid habit and slender twiggy stems bearing stiff, narrow, dark green, prickly leaves, to 1.2 cm x 0.2 cm, with the margins curved under. Clusters of tiny, white or pink, urn-shaped flowers, with spreading hairless lobes, are carried in the upper leaf axils and followed by small white fleshy fruit.

Flowering period	Aug. to Nov.
Distribution	Eastern suburbs, but generally uncommon, and scattered throughout the state; also Qld, NSW, ACT, Tas and SA.
Habitat	Gravelly or stony ridges and slopes in open forest.
Notes	This species is sometimes locally common, growing as scattered plants in loose colonies. The flowers have a honey scent.
Similar species	Could be mistaken for a *Leucopogon* but the flowers are hairless.
Cultivation	Requires well-drained soil but is slow growing and difficult to propagate.

Family Campanulaceae

Description

Perennial herb with erect to sprawling, angular or winged stems, rooting from the nodes, and clothed with soft, dark green leaves, to 7 cm x 1.7 cm, with entire or toothed margins. Blue to purplish, fan-shaped flowers, about 0.8 cm across, are borne in the upper leaf axils.

Flowering period	Mainly Nov. to May, but also sporadic at other times.
Distribution	Widespread in the Melbourne region and throughout the state; also Qld, NSW, Tas, SA, WA, NZ and South Africa.
Habitat	Moist areas in open forest, woodland, grassland, heathland, coastal scrubs and marshes.
Notes	This species may form dense clumps, spreading by self-layering stems.
Cultivation	Readily grown in moist to wet soils.

Family Lomandraceae

Description

Perennial herb, to 50 cm tall, forming sparse tussocks and sometimes spreading by underground stolons. The stiff, dull green to bluish-green, willowy leaves, to 50 cm x 0.3 cm, have inrolled margins, and are tipped with one to three tiny points. Small, yellow, tubular or globular, unisexual flowers are borne in slender branched or unbranched racemes, to 30 cm tall.

Flowering period	Sept. to Dec.
Distribution	Widespread in the Melbourne region and throughout the state; also Qld, NSW and ACT.
Habitat	Open forest and woodland.
Notes	Plants of this species really only become noticeable at flowering time.
Similar species	None in the region.
Cultivation	Requires excellent drainage. Suitable as a container plant or among rocks.

Family Lomandraceae

Description

Perennial herb forming dense tussocks, with numerous, flat, yellowish-green to dark green leaves, to 100 cm x 0.7 cm, which have two or three, prominent apical teeth. Small, tubular, unisexual flowers, yellowish with purplish bases, are borne in dense clusters on flat stalks, to about 80 cm tall. These are followed by hard, green to brown, shiny fruit.

Flowering period	Sept. to Dec.
Distribution	Widespread in the Melbourne region and throughout the state; also Qld, NSW, ACT, Tas and SA.
Habitat	Open forest, heathy forest, woodland and heathland.
Notes	This species is extremely variable in vigour, flower colour and the arrangement of the flowers in the inflorescence. The flowers of some variants have a strong chemical odour.
Similar species	None in the region.
Cultivation	Readily grown in a range of well-drained soils and positions.

Family Lomandraceae

Description

Perennial herb with sparse to dense tussocks of thick, leathery, grey to bluish-green leaves, to 90 cm x 0.3 cm. Small, bell-shaped, semi-pendent, unisexual, pinkish flowers are massed in whorls on prominent, erect inflorescences. Those bearing male flowers are branched, whereas the female flowers are carried on unbranched racemes.

Flowering period	June to Jan.
Distribution	Eastern suburbs, Dandenong Ranges, and throughout much of the state; also Qld, NSW, ACT, SA, WA and NT.
Habitat	Open forest, heathy forest, woodland and heathland.
Notes	The striking inflorescences of this species always attract attention.
Similar species	None in the region.
Cultivation	Requires excellent drainage.

Family Proteaceae

Description

Shrub, to 2 m tall, with a thick, woody rootstock, stiffly erect, slender stems, rust-coloured hairy new growth and leathery, dark green, holly-like leaves, to 18 cm x 5 cm, with a prominent network of veins. Creamy white to yellowish flowers, which are well-displayed on long racemes above the foliage, are followed by brown, papery capsules, to 3 cm long.

Flowering period	Dec. to Feb.
Distribution	Eastern suburbs, Dandenong Ranges, and mainly in southern areas of the state; also Qld and NSW.
Habitat	Moist forests, particularly on sheltered slopes and valleys.
Notes	Plants may be bushy or sparse, sometimes with only a few stems. Vigorous regrowth occurs after fire.
Similar species	None in the region.
Cultivation	Requires excellent drainage in filtered sun or semi-shade.

Family Proteaceae

Description

Bushy shrub, to 5 m tall, with reddish stems and slender, leathery leaves, to 20 cm x 2 cm, with a prominent pale midrib and coarsely toothed margins. Greenish-white or cream, perfumed flowers, which are carried in axillary racemes, to 15 cm long, are followed by dark brown, leathery capsules, about 2.5 cm long.

Flowering period	Dec. to Feb.
Distribution	*Rare in the region* where it is restricted to a couple of eastern suburbs, and in eastern parts of the state; also NSW, ACT.
Habitat	Moist forests near streams.
Notes	A distinctive species readily recognised by the long narrow toothed leaves.
Similar species	None in the region.
Cultivation	Readily grown in a moist sheltered position.

Family Orchidaceae

Description

Terrestrial orchid with a single, stiff leathery leaf, to 20 cm x
1.2 cm, which is dark green on the upper surface and greyish to
whitish beneath. A sturdy, wiry flower-stem, to 45 cm tall, carries
two to eight, strongly scented flowers, 2.5–3 cm across, with
narrow segments. These are usually dark brown or reddish-brown,
and the upper half of the lip is a contrasting bright yellow.

Flowering period Sept. to Nov.

Distribution   Eastern suburbs, Dandenong Ranges, and eastern and
                some western areas of the state; also Qld, NSW, ACT
                and Tas.

Habitat   Open forest, heathy forest and heathland.

Notes   Occurs mainly as scattered individuals or in loose
                colonies. Plants can be identified readily by the
                distinctive leaf.

Similar species None in the region.

Cultivation   Grown successfully in containers by orchid specialists.

Family Marsileaceae

Description

Small aquatic fern with clover-like fronds developing from a very slender creeping rhizome buried in mud. The fronds may float on water, or be erect in drying habitats. Each has four wedge-shaped, silky-haired or hairless segments arranged like a four-leaf clover at the end of a slender stalk. The spores are borne in hard, nut-like structures on slender stalks, which arise from the base of the frond stalk.

Distribution	Western and northern suburbs, and throughout western and northern areas of the state; also Qld, NSW, SA, WA and NT.
Habitat	Wet depressions, swamps, clay pans, waterholes and billabongs, in woodland and grassland.
Notes	Can grow as a submerged aquatic, emergent on swamp margins or in moist to wet grassland.
Similar species	*M. hirsuta* has very short stalks on the fruiting bodies.
Cultivation	Easily grown in ponds or wet soil.

Family Myrtaceae

Description

Dense shrub or tree, to 12 m tall, with whitish, papery bark and twiggy stems bearing very narrow, dark green, blunt leaves, to 1.5 cm long. White to cream, scented flowers are borne in dense, well-displayed terminal spikes, to 2.5 cm x 1.5 cm, and are followed by clusters of grey, woody capsules.

Flowering period	Oct. to Dec.
Distribution	Widespread in the Melbourne region and in eastern parts of the state; also Qld, NSW and Tas.
Habitat	This species often grows in dense suckering colonies to the exclusion of most other plants.
Similar species	*M. parvistaminea* has smaller leaves and narrower flower spikes.
Cultivation	Easily grown in moist to wet soils.

Family Myrtaceae

Description

Dense shrub or small tree, to 10 m tall, with a gnarled trunk and a widely spreading bushy canopy. Twiggy stems bear stiff, narrow, dark green, pointed leaves, to 1.2 cm x 0.1 cm. White or cream flowers are well-displayed in dense terminal spikes, to 5 cm x 1.5 cm, and are followed by clusters of grey, woody capsules.

Flowering period	Oct. to March
Distribution	Western and eastern bayside suburbs, and over much of the state; also Qld, NSW, SA and WA.
Habitat	Coastal scrubs, coastal headlands and streamside vegetation.
Notes	This species, which withstands considerable coastal exposure, is also distributed further inland where it often grows in heavy soils.
Similar species	None in the region.
Cultivation	Widely grown in windbreaks and shelter belts. Very adaptable.

173

Family Myrtaceae

Description

Dense shrub or small tree, to about 10 m tall, with cream, papery bark and bright green young growth. Twiggy stems are clothed with stiff, dark green, pleasantly aromatic leaves, to 1.5 cm x 0.7 cm, arranged in overlapping opposite pairs. Cream to yellow, fragrant flowers are well-displayed in dense, terminal spikes, to 5 cm x 1.5 cm, and are followed by clusters of woody capsules.

Flowering period	Sept. to Feb.
Distribution	Eastern and south-eastern suburbs, and in much of the state; also NSW, Tas and SA.
Habitat	Swamps and wet peaty soil.
Notes	This species sometimes grows in dense thickets to the exclusion of nearly all other plants. The leaves have a pleasant fragrance when crushed.
Similar species	None in the region.
Cultivation	Readily grown in moist soils in positions exposed to some sun.

Family Poaceae

Description

Perennial grass forming dense tussocks of crowded, bright green
leaves, to 8 cm x 0.2 cm, and erect flower-stems. Small, greenish
flowers with long, prominent awns are borne in slender uncrowded
panicles, which weep gracefully at the top.

Flowering period	Mainly Sept. to Dec., but also sporadic after rain.
Distribution	Widely distributed in the Melbourne region and throughout the state; occurs in all states and NZ.
Habitat	Open forest, heathy forest, woodland and grassland.
Notes	An extremely important native grass. A variable species with some forms remaining in tight clumps and others spreading in swards. Recent studies suggest it should be included in the genus *Ehrharta*.
Similar species	None in the region.
Cultivation	Readily grown and excellent in shady positions.

175

Family Asteraceae

Description

Tufted herb with a carrot-like rootstock and a rosette of prostrate to erect, dark green, shiny leaves, which have a few coarse marginal teeth. Buds nod or droop only becoming erect just before opening. Bright yellow, daisy flower-heads, 3–4 cm across, arise singly on slender stalks, to 50 cm long, and are followed by fluffy, white seed-heads.

Flowering period	July to Jan.
Distribution	Mainly in the eastern suburbs, Dandenong Ranges, and very widely distributed in the state; also NSW, ACT, Tas, SA and WA.
Habitat	Grassland and grassy areas in open forest and woodland.
Notes	A showy native herb which has a similar general appearance to the introduced weedy dandelion. The roots of this plant were eaten by the Aborigines.
Similar species	Dandelions have erect buds.
Cultivation	An attractive subject for containers.

Family Orchidaceae

Description

Slender terrestrial orchid with a hollow, green, onion-like leaf, to 35 cm long, and a flower-stem, to 40 cm tall, which carries a spire-shaped spike of numerous, crowded, green to yellowish green flowers, about 0.3 cm across. These flowers sit atop a prominent swollen ovary.

Flowering period	Oct. to Dec.
Distribution	Northern and eastern suburbs and throughout much of the state; also Qld, NSW, ACT, Tas and SA.
Habitat	Grassy areas in open forest, heathy forest, woodland and grassland.
Notes	A weedy orchid which can colonise suitable moist grassy sites, including lawns and pasture.
Similar species	*M. unifolia* has a notch on the apex of the lip.
Cultivation	Readily grown in a container and can be established in a sheltered position in the garden.

Family Asteraceae

Description

Perennial herb often forming a compact mound, with twiggy stems clothed in narrow, bright green leaves to 2.5 cm long. Each stem ends in a daisy flower-head, about 2 cm across, which is usually white, but may also be lilac, mauve or purplish.

Flowering period	Mainly Aug. to Feb., but also sporadic at other times.
Distribution	Western and northern suburbs, and in northern parts of the state; also Qld, NSW, SA, WA and NT.
Habitat	Open grassland and often also colonising road verges.
Notes	Plants frequently become smothered in flowers and provide an attractive display.
Similar species	None in the region.
Cultivation	Requires very well-drained soil in a sunny position.

Family Myoporaceae

Description

Robust shrub, to 6 m tall, with a bushy, rounded to spreading canopy and shiny green, sometimes sticky, new growth. Thick, dark green, fleshy leaves, to 8 cm x 2.5 cm, clothe the brittle branches. White flowers, about 1 cm across, which are borne in axillary groups of up to eight, are followed by globular, succulent fruit, about 0.8 cm across, which are white with purple markings when ripe.

Flowering period	Mainly May to Nov., but also sporadic at other times.
Distribution	Western suburbs, eastern bayside suburbs, and over southern parts of the state; also NSW, SA and WA.
Habitat	Coastal scrub and coastal headlands.
Notes	An important species for stabilising coastal dunes and headlands. The fruit are eaten by birds.
Similar species	*M. viscosum* has leaves with conspicuously toothed margins.
Cultivation	Adaptable to a wide range of positions but excellent for coastal districts.

Family Haloragaceae

Description

Robust aquatic which forms dense clumps with two distinct leaf shapes, both types in whorls of five to eight leaves. Submerged stems have pinnate leaves with thread-like lobes, whereas the erect stems have pinnate leaves with much broader segments, and some of the upper leaves can even be undivided.

Flowering period	Oct. to April.
Distribution	Western and eastern suburbs, and widespread throughout the state; also Qld, NSW, Tas, SA and WA.
Habitat	Wet depressions, swamps, billabongs and slow-moving streams.
Notes	The emergent leafy stems of this species look remarkably like miniature pine trees.
Cultivation	Readily grown in ponds or dams or even in wet boggy sites.

Family Asteraceae

Description

Shrub or small tree, to 8 m tall, with a gnarled trunk covered with ribbony grey bark, spreading canopy, and silvery new growth. Large, leathery leaves, to 20 cm x 9 cm, are dark green and shiny above, contrasting markedly with the silvery-white lower surface. White and yellow, daisy flower-heads, 2 x 2.5 cm across, are borne in large, conspicuous terminal clusters.

Flowering period	Oct. to Dec.
Distribution	Eastern suburbs, Dandenong Ranges, and widespread in cooler parts of the state; also NSW, ACT and Tas.
Habitat	Moist sheltered slopes and gullies in tall forest.
Notes	This species often grows in extensive stands. The contrasting leaf surfaces are attractive in windy weather.
Similar species	None in the region.
Cultivation	Best in a sheltered position.

Family Asteraceae

Description

Dense, silvery-grey shrub, to 2 m tall, with a bushy to spreading habit and woody stems densely clothed with silvery-grey to whitish, hairy, narrow leaves, to 4.5 cm x 0.5 cm. Yellowish flower-heads, borne in the upper axils of short side growths, are rather insignificant.

Flowering period	Jan. to April.
Distribution	Eastern and south-eastern bayside suburbs, and mainly in coastal regions of the state; also NSW, Tas, SA and WA.
Habitat	Coastal scrub and coastal headlands.
Notes	An important component of coastal vegetation for stabilising sand dunes. Although the flowers are insignificant the foliage of this species is highly decorative.
Similar species	None in the region.
Cultivation	Excellent for coastal districts. Requires well-drained soil in a sunny position.

Family Asteraceae

Description

Sparse shrub, to 1.5 m tall, with slender, erect to spreading branches
clothed with dark green, shiny leaves, to 2 cm x 0.4 cm, which are
whitish beneath. Unusual flower-heads, about 2 cm across, which
have a yellow to bluish centre, and two to four spreading white rays,
are borne in clusters near the end of the branches.

Flowering period	Oct. to Feb.
Distribution	Eastern suburbs, Dandenong Ranges, and throughout much of the state; also NSW, ACT and Tas.
Habitat	Hilly districts, often among rocks in open forest.
Notes	Plants sometimes flower profusely and can exude a coconut-like scent.
Similar species	*O. erubescens* has larger flower-heads with 4 to 7 spreading rays.
Cultivation	Grows well in a moist sheltered position.

Family Asteraceae

Description

Shrub, to 3 m tall, with an open to dense habit and pithy stems clothed with soft, greyish-green leaves, to 8 cm x 1 cm, which are whitish beneath. Daisy flower-heads, about 1.5 cm across, which are borne in large clusters from the upper leaf axils, are commonly white, but can also be blue, mauve or pink.

Flowering period	Aug. to Jan.
Distribution	Eastern suburbs, and widespread in eastern parts of the state; also NSW, ACT and Tas.
Habitat	Moist sheltered sites in open forest.
Notes	A fast-growing species which may be prominent after fire. The flower-heads are well-displayed and the variation in colour provides scope for selection of variants for cultivation.
Similar species	None in the region.
Cultivation	An excellent species which has proved to be very adaptable. Best if pruned heavily after flowering.

Family Ophioglossaceae

Description

Small fern which grows in compact groups or colonies, spreading by slender underground stolons. The sterile fronds consist of an undivided, ovate, green blade, to 5 cm x 1.5 cm. The fertile frond, which arises from the base of the sterile frond, consists of a slender spore-bearing spike atop a slender stalk to 15 cm tall.

Distribution	Northern and eastern suburbs, and throughout much of the state; occurs in all states and many overseas countries.
Habitat	Grassland, heathland, open forest and woodland, sometimes in shallow soil over rock sheets.
Notes	Often only the simple sterile fronds can be found. Previously known as *O. coriaceum*.
Similar species	None in the region.
Cultivation	Easily grown in a container but difficult to establish in the garden.

Family Orchidaceae

Description

Terrestrial orchid with a basal tuft of two to five, narrow, grass-like leaves, to 30 cm x 0.3 cm, and a sturdy, rigid flower-stem, to 60 cm tall, which carries one to nine flowers, about 1 cm across. These are dark brown to blackish with a prominent yellow patch on the lip, and two unusual, horn-like sepals which extend above the central part of the flower.

Flowering period	Nov. to Jan.
Distribution	Eastern suburbs, Dandenong Ranges, and scattered mainly in southern parts of the state; also Qld, NSW, Tas, SA, NZ and NCal.
Habitat	Open forest, heathy forest and heathland, often in moist depressions and near swamps.
Notes	This orchid is not conspicuous, it is difficult to see among other plants.
Similar species	None in the region.
Cultivation	There has been little success in cultivating this plant.

Family Asteraceae

Description

Shrub, to 3 m tall, with a bushy rounded habit, yellowish new growth and densely woolly branchlets, bearing stiff, narrow leaves, to 2.5 cm x 0.2 cm, which are green on the upper surface and woolly-white beneath. Small, cream to yellowish flower-heads are crowded in conspicuous terminal clusters.

Flowering period	Feb. to May.
Distribution	Eastern and south-eastern bayside suburbs, and coastal districts of the state; also Qld, NSW, Tas and SA.
Habitat	Coastal scrub and coastal headlands.
Notes	This species tolerates considerable exposure to buffeting salt-laden winds.
Similar species	None in the region.
Cultivation	Requires excellent drainage and regular pruning. Excellent for exposed coastal situations.

Family Bignoniaceae

Description

Robust high-climbing twiner, which becomes densely bushy in exposed situations, with dark, shiny new growth and large pinnate leaves. These have dark green shiny leaflets, to 5 cm x 1 cm, with the apical leaflet longest. Tubular, creamy white flowers, to 3 cm long, in large terminal bunches, have prominent maroon to purplish-red markings in the throat.

Flowering period	Sept. to Jan.
Distribution	Eastern suburbs, Dandenong Ranges, and throughout much of the state; also Qld, NSW and Tas.
Habitat	Moist sheltered valleys in tall forest.
Notes	This species is very attractive when in flower.
Similar species	None in the region.
Cultivation	Readily grown in a moist sheltered position.

Family Iridaceae

Description

Small, perennial, iris-like herb with tufts or fans of stiff, dull, bluish-green leaves, to 20 cm long, and flower-stems much shorter than the leaves, ending in large leathery bracts. Bright blue to deep purple, iris-like flowers are produced at intervals from the bracts. Each is dominated by three large, flimsy segments.

Flowering period	Sept. to Jan.
Distribution	Eastern bayside suburbs, and widespread in near-coastal areas of the state; also Qld, NSW, Tas and SA.
Habitat	Heathland and heathy forest.
Notes	In some areas, plants of this species have terete leaves, whereas they are flat in other sites.
Similar species	*P. occidentalis* has flower spikes that are much longer than the leaves.
Cultivation	Readily grown in a container or among rocks in the garden.

Family Geraniaceae

Description

Perennial herb with a thick carrot-like rootstock and an open rosette of long-stalked, dark green leaves, to 4 cm x 3 cm, with prominently toothed or lobed margins. Bright magenta-pink flowers with darker veins, to 2.5 cm across, are borne in clusters on the end of slender leafless stalks, to 10 cm long.

Flowering period	Nov. to March.
Distribution	Northern suburbs, where sporadic, and throughout much of the state; also NSW, SA and WA.
Habitat	Among rocks in grassland.
Notes	A very showy species which sometimes grows in localised colonies.
Similar species	None in the region.
Cultivation	An excellent rockery plant which needs well-drained soil.

Family Sinopteridaceae

Description

Ground fern forming spreading patches, with slender subterranean rhizomes and erect, fishbone-like, dark green, shiny fronds, to 80 cm tall. Each frond has narrow, spreading, straight or curved, leathery segments, to 5 cm x 1 cm. The spores are borne in brown marginal bands.

Distribution	Widespread in the Melbourne region and throughout much of the state; also Qld, NSW, ACT, Tas, NZ and NCal.
Habitat	Among rocks in open forest and woodland.
Notes	This species forms spreading patches. In dry times the fronds may curl and shrivel, but revive after rain.
Similar species	None in the region.
Cultivation	Readily grown in well-drained soil in partial or filtered sun.

Family Polygonaceae

Description

Perennial herb forming spreading clumps, with fleshy, prostrate stems clothed with yellowish-green to bright green leaves, to 15 cm x 1 cm, which have brown markings. Tiny pink flowers are clustered on slender spikes, to 10 cm long, which arise from the end of branches.

Flowering period	Sporadic all year.
Distribution	Widespread in the Melbourne region and throughout the state; occurs in all states.
Habitat	Moist to wet depressions and swamps in open forest and woodland.
Notes	This species may grow as an aquatic. The stems take root as they spread.
Similar species	*P. hydropiper* has green to white flowers on pendulous spikes.
Cultivation	Readily grown in ponds or boggy areas.

Family Thymeleaceae

Description

Slender, few-stemmed shrub, to 2 m tall, with erect, willowy, reddish branches clothed with crowded, short, broad, yellowish-green leaves, to 2 cm x 1 cm, in opposite pairs. Small, bright yellow flowers are well-displayed in conspicuous terminal heads, 1–1.5 cm across.

Flowering period	Aug. to Dec.
Distribution	Eastern suburbs, Dandenong Ranges, and widespread in southern parts of the state; also Tas and SA.
Habitat	Moist forests.
Notes	This species regenerates strongly after fire. The stems are covered with tough pliant bark which can be used as string.
Similar species	None in the region.
Cultivation	Requires a sheltered position and summer moisture. Responds well to regular pruning.

Family Thymeleaceae

Description

Dwarf shrub, to 0.5 m tall, which forms a low, straggly to mounded clump, sometimes spreading by suckers. The slender, hairy branchlets carry opposite pairs of narrow, green to grey green leaves, about 1.5 cm x 0.5 cm, and end in showy clusters, 2–2.5 cm across, of creamy white flowers.

Flowering period	Sept. to Jan.
Distribution	Widespread in the Melbourne region and throughout much of the state; also NSW, Tas and SA.
Habitat	Open forest, woodland, heathy forest, heathland and grassland.
Notes	A showy species which attracts butterflies when in flower.
Similar species	*P. glauca* has hairless branchlets.
Cultivation	Readily grown in containers or in the garden. Looks attractive in a rockery.

Family Fabaceae

Description

Erect or scrambling shrub, to 1.5 m tall, with thin, wiry branches and opposite pairs of stalkless, dark green, stiff, leathery, heart-shaped leaves, to 9 cm x 4 cm, with prominent veins. Showy, red and yellow, pea-shaped flowers, about 2 cm across, are borne on long, hairy stalks in the axils, and are followed by flat, leathery pods, to 5 cm x 2 cm, which split open to reveal a satiny interior.

Flowering period	Sept. to Dec.
Distribution	Eastern and south-eastern suburbs, Dandenong Ranges, and throughout much of the state; also Qld, NSW, ACT and Tas.
Habitat	Moist forests, heathy forest and heathland.
Notes	This species may have trailing stems or scramble through surrounding shrubs. The flowers open only on sunny days.
Similar species	*P. obtusangulum* has triangular leaves with sharp points at the angles.
Cultivation	Requires well-drained soil in a sheltered position.

Family Asteraceae

Description

Perennial herb with a thin rootstock and rosette of dark green leaves, to 20 cm x 2 cm, from which arise numerous, slender, reddish, leafy stems. Each stem bears one to a few, daisy-like flower-heads, to 3 cm across, which are nodding when in bud, and open bright yellow, with a dense cushion of central florets and a few flat rays spreading from the margins.

Flowering period	Oct. to Dec.
Distribution	Widespread in the Melbourne region and throughout much of the state; also Qld, NSW, ACT, Tas and SA.
Habitat	Grassland and grassy areas of open forest and woodland.
Notes	This species is spectacular when in flower.
Similar species	None in the region.
Cultivation	An excellent plant for growing in containers or among rocks.

Family Dryopteridaceae

Description

Ground fern growing in neat, discrete clumps, with an erect rootstock, sometimes forming a short trunk, and a crown of arching, dark green fronds, to 1.5 m x 30 cm. Each frond has a short, densely scaly stalk and an oblong blade, divided two or three times, which is rough to the touch, and often bears bulbils or plantlets towards the tip. New fronds are pale green. The spores are borne in small circular clusters.

Distribution	Eastern suburbs, Dandenong Ranges, and widespread in southern areas of the state; also NSW, ACT and Tas.
Habitat	Moist sheltered slopes and gullies in tall forest.
Notes	This species, which sometimes grows in extensive colonies, reproduces commonly by the plantlets which develop on the older fronds.
Similar species	None in the region.
Cultivation	A very hardy and adaptable fern.

Family Rhamnaceae

Description

Shrub or small tree, to 10 m tall, with rust-coloured, hairy new growth and large, soft, dark green leaves, to 14 cm x 6 cm, with prominent veins. Small, cream to green flowers are generally inconspicuous, but are carried in large, branched sprays at the end of the shoots.

Flowering period	Oct. to Dec.
Distribution	Northern and eastern suburbs, Dandenongs, and throughout cooler parts of the state; also Qld, NSW, ACT, Tas and SA.
Habitat	Moist sheltered slopes and valleys, especially near streams.
Notes	The flowers of this species lack petals. Trunks of larger plants often support colonies of lichen.
Similar species	None in the region.
Cultivation	A fast-growing species for a moist sheltered situation.

Family Potamogetonaceae

Description

Aquatic plant with fleshy stems rooting in the mud, and bearing floating leaves which are attached to the stems by long stalks. The leaves, which are mostly ovate to oblong in shape, to 10 cm x 2.5 cm, are dark green, thick and fleshy. Small greenish flowers are borne in dense spikes, to about 4 cm long, at the end of long stalks.

Flowering period	Sept. to April.
Distribution	Northern and eastern suburbs, and throughout the state; occurs in all states.
Habitat	Swamps, billabongs, dams and slow-moving streams.
Notes	In suitable sites this species can form dense patches.
Cultivation	Readily grown in ponds and dams.

Family Orchidaceae

Description

Slender terrestrial orchid with a hollow onion-like leaf, to 35 cm long, green with a red base, and a flower-stem, to 90 cm tall, with a densely crowded spike of 15–60, upside-down flowers, each about 1.5 cm across. These are yellowish-green to brownish, with reddish stripes and a prominent crystalline-white lip, with frilly margins.

Flowering period	Nov. to Dec.
Distribution	Eastern suburbs, Dandenong Ranges, and in southern parts of the state; also Qld, NSW, Tas and SA.
Habitat	Moist depressions and swamps in open forest, heathy forest and heathland.
Notes	This species mainly grows in small colonies. Flowering is strongly promoted by summer fires and the flowers are sweetly perfumed in warm weather.
Similar species	None in the region.
Cultivation	There has been little success in cultivating this plant.

Family Orchidaceae

Description

Slender terrestrial orchid with a hollow onion-like leaf, to 35 cm long, green or blackish with a reddish base, and a flower-stem, to 50 cm tall, with a spike of five to 30, upside-down flowers, each about 1 cm across. These are green to brownish, with a conspicuous white lip which is doubled back on itself.

Flowering period	Oct. to Dec.
Distribution	Eastern suburbs, Dandenong Ranges, and mainly in eastern parts of the state; also Qld, NSW, ACT and Tas.
Habitat	Open forest, heathy forest, woodland and heathland.
Notes	Flowering in this species is dramatically promoted by summer fires. The flowers lack any scent.
Similar species	*P. odoratum* has fragrant flowers and the lip curves back through the sepals.
Cultivation	There has been little success in cultivating this plant.

Family Orchidaceae

Description

Slender terrestrial orchid with a hollow, green to blackish, onion-like leaf, to 120 cm tall, and a flower-stem, to 150 cm tall, with a spike of 15–60, upside-down flowers, each about 1.5 cm across. These are variable in colour, from yellowish-green to purplish-black, with a conspicuous crinkly, white or pink lip and a yellow callus.

Flowering period	Oct. to Dec.
Distribution	South-eastern suburbs, and in southern parts of the state; also Qld, NSW, Tas, SA and WA.
Habitat	Open forest, heathy forest and heathland.
Notes	Flowering in this species is dramatically promoted by summer fires. The flowers are fragrant in warm weather.
Similar species	None in the region.
Cultivation	There has been little success in cultivating this orchid.

Family Lamiaceae

Description

Perennial herb which forms rounded to spreading clumps, usually with an open appearance. Fleshy, reddish stems, have opposite pairs of dark green, hairy, ovate leaves, to 6 cm x 2 cm, and end in dense, compact heads of purplish-brown bracts, from which arise purple, tubular, two-lipped flowers.

Flowering period	Mainly Nov. to Feb., but also sporadic at other times.
Distribution	Northern and eastern suburbs, and widespread throughout the state; also Qld, NSW, ACT, Tas, SA and NZ.
Habitat	Moist to semi-swampy areas of open forest, woodland and heathland.
Notes	A fast growing species which may colonise disturbed sites in wet areas.
Similar species	None in the region.
Cultivation	Readily grown in moist soils.

Family Dennstaedtiaceae

Description

Coarse ground fern forming widely spreading patches, with vigorous coarse subterranean rhizomes and dark green, leathery fronds, to 1.5 m x 1 m. Each frond has a long stout brown stalk and a triangular blade, divided two or three times, which has numerous narrow segments. The spores are borne in narrow marginal bands.

Distribution	Widespread in the Melbourne region and throughout most of the state; also Qld, NSW, ACT, Tas, SA, NZ and many other countries.
Habitat	Well-drained soils in most habitats.
Notes	A robust invasive fern that is poisonous to stock.
Similar species	None in the region.
Cultivation	Resents disturbance and can be very difficult to establish.

Family Orchidaceae

Description

Slender terrestrial orchid with three to five, dark green, succulent leaves, to 6 cm x 1.5 cm, scattered up a fleshy flower-stem, the basal leaves being largest. Plants have a single hooded flower, about 3 cm long, which is translucent white, with dark green lines and markings. The top of the flower curves forwards and the sepal tips sweep back strongly. The curved lip protrudes prominently from the front of the flower.

Flowering period	Sept. and Oct.
Distribution	Eastern suburbs, Dandenong Ranges, and mainly in eastern parts of the state; also NSW and ACT.
Habitat	Moist slopes and gullies in tall forest.
Notes	A distinctive species which grows in small loose colonies.
Similar species	*P. falcata* has much larger sickle-shaped flowers.
Cultivation	Grown successfully in containers by orchid specialists.

Family Orchidaceae

Description

Small terrestrial orchid with a basal rosette of four to six, dark green, short-stalked leaves, to 3 cm x 1.5 cm, which often have wavy margins. A slender central flower-stem, to 30 cm tall, bears a single hooded flower, about 1.5 cm long. This is a mixture of translucent white, dark green and reddish-brown, with the sepal tips extending above the top of the flower. The apex of the lip is broadly notched.

Flowering period	June to Sept.
Distribution	Eastern and south-eastern suburbs, and widespread in southern parts of the state; also NSW, Tas and SA.
Habitat	Open forest, heathy forest and heathland.
Notes	This species forms extensive crowded colonies and flowers freely.
Similar species	None in the region.
Cultivation	Grown successfully in containers by specialist growers.

Family Orchidaceae

Description

Terrestrial orchid with three to seven, dark green, succulent leaves, to 10 cm x 3 cm, scattered up a fleshy flower-stem, the upper one often sheathing the base of the flower. Plants have a single hooded flower, about 4 cm long, which is translucent white and green, with the front sepals rich reddish-brown. The lip is just visible through the notch at the front of the flower.

Flowering period	Sept. to Nov.
Distribution	Eastern bayside suburbs, but *now extinct there,* although still known in the Mornington Peninsula; *also rare in other parts of the state, and in Tas and SA.*
Habitat	Tea-tree heath on stabilised dunes.
Notes	A majestic orchid which grows in colonies.
Cultivation	Grown successfully in containers by orchid specialists.

207

Family Orchidaceae

Description

Terrestrial orchid with a basal rosette of three to six, dark green,
stalked leaves, to 10 cm x 3 cm, which often have wavy margins. A
central flower-stem, to 30 cm tall, bears a single hooded flower, to
3.5 cm long. This is green and white with brown shading towards
the apex. The apex of the lip is characteristically twisted to one side.

Flowering period	Aug. to Oct.
Distribution	Eastern suburbs, and widespread throughout much of the state; also Qld, NSW, ACT, Tas and SA.
Habitat	Moist areas in open forest, woodland and heathy forest.
Notes	This species grows in extensive dense colonies and flowers freely.
Similar species	None in the region.
Cultivation	Grown successfully in containers by specialist growers.

Family Orchidaceae

Description

Terrestrial orchid with a basal rosette of three to six, dark green, stalked leaves, to 9 cm x 3 cm, which often have wavy or crinkled margins. A central flower-stem, to 30 cm tall, bears a single, nodding, hooded flower, to 2.5 cm long. This is light translucent green with darker green shading towards the apex. The hairy lip protrudes prominently from the flower.

Flowering period	July to Oct.
Distribution	Northern and eastern suburbs, Dandenong Ranges, and widespread over much of the state; also Qld, NSW, ACT, Tas, SA and NZ.
Habitat	Moist areas in open forest, woodland and heathy forest.
Notes	This familiar orchid grows in extensive, often dense colonies and flowers freely.
Similar species	None in the region.
Cultivation	Grown successfully in containers by specialist growers.

Family Orchidaceae

Description

Terrestrial orchid with a basal rosette of four to six, dark green, stalked leaves, to 4 cm x 2 cm, which often have wavy margins. A central flower-stem, to 25 cm tall, bears a single flower, to 2 cm long. This is green and white at the base, and dark reddish-brown or purplish-brown in the front and towards the apex. The tip of the lip is just visible from the front of the flower.

Flowering period	Aug. to Oct.
Distribution	Northern and eastern suburbs, Dandenong Ranges, and throughout much of the state; also NSW, ACT, Tas and SA.
Habitat	Moist sites in open forest, tall forest, heathy forest and coastal scrub.
Notes	This species grows in extensive, often dense colonies and flowers freely.
Similar species	None in the region.
Cultivation	Grown successfully in containers by specialist growers.

Family Amaranthaceae

Description

Perennial herb with a thick carrot-like rootstock and a basal rosette of dark green, narrow leaves, to 10 cm x 0.5 cm, with wavy margins. Robust stems, to 100 cm tall, carry a single, fluffy or feathery terminal head, to 12 cm x 6 cm, which is composed of numerous, densely crowded, greyish-green to greenish-yellow flowers.

Flowering period	Oct. to Dec. and March to May.
Distribution	Western and northern suburbs, *but now very rare*, and mainly in southern areas of the state; also Qld, NSW, SA, WA and NT.
Habitat	Grassland and grassy woodland.
Notes	This species often occurs in heavy soils in open, sparsely covered sites or among rocks.
Similar species	None in the region.
Cultivation	An interesting container plant.

Family Fabaceae

Description

Shrub, to about 3 m tall, with a dense bushy habit, light green new growth, and distinctive, wedge-shaped leaves, to 3 cm x 1 cm, which are dark olive green above and paler beneath. Bright yellow and red-brown, pea-shaped flowers, about 1 cm across, which are carried in prominent, dense terminal clusters, are followed by small brown pods.

Flowering period	Aug. to Nov.
Distribution	Eastern and south-eastern suburbs, Dandenong Ranges, and widespread throughout the state; also NSW, Tas and SA.
Habitat	Open forest and woodland.
Notes	This showy species is sometimes prominent on disturbed sites such as road verges.
Similar species	None in the region.
Cultivation	Fast growing, but requires excellent drainage.

Family Fabaceae

Description

Shrub, to about 1.5 m tall, with slender, wiry, erect to spreading branches clothed with tiny, ovate, dark green leaves, to 0.6 cm long, which have the margins curved under. Bright yellow, orange and red, pea-shaped flowers, about 1 cm across, are carried in prominent terminal clusters, and are followed by small, inflated, hairy pods.

Flowering period	Sept. to Nov.
Distribution	Eastern suburbs, Dandenong Ranges, and mainly in eastern parts of the state; also Tas.
Habitat	Moist slopes and valleys in open forest and woodland.
Notes	This species can produce spectacular floral displays, especially in the first three to five years after a fire.
Similar species	*P. stricta* has leaves to 1 cm long.
Cultivation	Fast growing and adaptable, but requires excellent drainage.

Family Orchidaceae

Description

Terrestrial orchid, with a thick, fleshy, dark green, ovate, ground-hugging leaf, to 10 cm x 8 cm, and a flower-stem, to 25 cm tall, with several large bracts and two to ten flowers, 3.5–4 cm across. Each flower has a large, hooded, white and red-striped rear sepal, with the other segments being narrow and dark reddish-brown. The lip has deeply fringed margins.

Flowering period	Sept. to Nov.
Distribution	Eastern and south-eastern suburbs, and scattered mainly in southern areas of the state; also NSW, Tas, SA and WA.
Habitat	Open forest, heathy woodlands and heathland.
Notes	The flowering of this species, which grows in extensive colonies, is strongly promoted by fire. Previously well known as *Lyperanthus nigricans.*
Similar species	None in the region.
Cultivation	There has been little success in cultivating this orchid.

Family Ranunculaceae

Description

Perennial herb with a rosette of soft, hairy, dark green leaves, each consisting of a long slender stalk and a lobed or deeply dissected blade, to 8 cm x 2.5 cm. Slender, branched flower-stems, to 50 cm tall, bear bright golden-yellow flowers, 2–2.5 cm across, on long thin stalks.

Flowering period	Mainly Sept. to Dec., but also sporadic at other times.
Distribution	Widespread in the Melbourne region and throughout the state; also Qld, NSW, ACT, Tas and SA.
Habitat	Moist grassy areas in grassland, open forest and woodland.
Notes	A distinctive species which often grows in small colonies. The seeds have an unusual coiled beak at one end.
Similar species	*R. pachycarpus* has less-divided leaves and sparsely branched flower-stems.
Cultivation	Readily grown in moist soil.

Family Chenopodiaceae

Description

Dense shrub with a widely sprawling or scrambling habit, and somewhat brittle branches, densely clothed with fleshy, dark green, succulent leaves, to 3.5 cm x 1 cm, which are paler beneath. Tiny whitish flowers, borne in dense pyramidal sprays, are unisexual, and the female flowers develop into flattish, shiny, dark red, succulent berries.

Flowering period	Dec. to April.
Distribution	Eastern and south-eastern bayside suburbs, and in southern parts of the state; also NSW, Tas, SA and WA.
Habitat	Coastal scrubs, coastal headlands and saltmarshes.
Notes	This species can climb strongly through surrounding vegetation. The fruit are eaten avidly by birds and were also a source of food for the Aborigines.
Similar species	*Einadia nutans* has greyish green leaves.
Cultivation	Readily grown in well-drained soil. Excellent for coastal districts.

Family Asteraceae

Description

Perennial herb with a woody rootstock and a sparse to compact, bushy habit, with short to long stems densely clothed with narrow, bluish-green, blunt leaves, to 1.8 cm x 0.2 cm. Slender stems, to 40 cm tall, carry single papery flower-heads, 2 x 3 cm across, which are white with a yellow centre.

Flowering period	Oct. to May.
Distribution	Western and northern suburbs, but *now scattered and rare*; widespread in the state, also Qld, NSW, ACT, Tas and SA.
Habitat	Grassland, sometimes among rocks and shrubs.
Notes	This species is variable in growth habit, leaf colour, and the size of the flower-heads. The buds are often pink to reddish. Previously well known as *Helipterum anthemoides*.
Similar species	None in the region.
Cultivation	Requires excellent drainage in a sunny location. Attractive when planted among rocks.

Family Euphorbiaceae

Description

Shrub, to 3 m tall, with a rounded to spreading, sparse or dense habit and twiggy branches clothed with narrow, dull-green or bronze leaves, to 4 cm long, with the margins rolled under. Showy white flowers, to 2.5 cm across, are produced in masses from the end of the branchlets. These flowers, which are males, surround an inconspicuous conical female flower which develops into a round fruit.

Flowering period	Mainly Sept. to Nov., but also sporadic at other times.
Distribution	South-eastern suburbs, *but now uncommon to rare*; also in eastern parts of the state, Qld, NSW and Tas.
Habitat	Heathland and heathy forest.
Notes	This species produces spectacular floral displays. The male flowers are fragrant.
Similar species	None in the region.
Cultivation	Requires excellent drainage in a sunny position. Very difficult to propagate.

Family Rosaceae

Description

Scrambling climber with slender, green to reddish stems covered with sharp, hooked thorns and bearing pinnate leaves, with three to five leaflets. These are light green above, silvery beneath, and with irregularly toothed margins. Small flowers, about 1.5 cm across, with large green sepals and deep pink petals, are followed by small, fleshy red berries.

Flowering period	Oct. to Dec.
Distribution	Widespread in the Melbourne region and throughout much of the state; also Qld, NSW, ACT, Tas and SA.
Habitat	Moist slopes and valleys in open forest and woodland.
Notes	The stems of this species may trail or climb weakly through surrounding vegetation. The fruit are edible, but without much flavour.
Similar species	None in the region.
Cultivation	Readily grown in a moist sheltered position.

Family Dryopteridaceae

Description

Epiphytic fern growing in slowly spreading clumps, with creeping, densely scaly, coarse rhizomes and bright green to yellowish-green, shiny, leathery fronds. Each frond has a long yellowish stalk and a triangular blade, to 40–50 cm in length, which is divided two to four times. The spores are borne in small circular clusters.

Distribution	Dandenong Ranges, and distributed disjunctly in southern areas of the state; also Qld, NSW, ACT, Tas, NZ and many other overseas countries.
Habitat	Moist, humid fern gullies.
Notes	Grows on tree fern trunks, butts of large trees, rotting logs and occasionally also on the ground.
Similar species	None in the region.
Cultivation	An excellent fern for hanging containers; requires shady, humid conditions and regular watering in the growing season.

Family Asteraceae

Description

Perennial herb which forms sparse to dense, spreading clumps with stems branching from near ground level, and narrow, dark green leaves, to 2 cm long, with the margins curved under. Bright yellow, button-like flower-heads, to 2 cm across, grow at the end of branches and are followed by grey fluffy seed-heads.

Flowering period	Nov. to May.
Distribution	Western suburbs, *but now very rare;* also NSW and ACT.
Habitat	Well-drained to moist sites in grassland.
Notes	This species, which has suffered greatly from clearing and urbanisation, is now regarded as being *endangered nationally.*
Similar species	*Leptorhynchos* species have similar flower-heads carried on thin wiry stems.
Cultivation	Readily grown in well-drained soil in a sunny location.

Family Orchidaceae

Description

Small epiphytic orchid consisting of a fan-like group of thin, flat, leathery, dark green leaves, to 8 cm x 1.4 cm, attached to a short stem, and with numerous thin, wiry roots. Pendulous flower-stems carry up to 14 fragrant flowers, each about 1.5 cm across, which are green to brownish with a prominent, purple-striped, white lip.

Flowering period	Nov. to Dec.
Distribution	Dandenong Ranges, and southern parts of the state, mainly in the east; also NSW and Tas.
Habitat	Moist slopes and gullies in tall forest.
Notes	This species grows on a wide range of host plants. It was once relatively common in the Dandenong Ranges, but has suffered greatly from the ravages of bushfires, urbanisation and illegal collecting.
Similar species	None in the region.
Cultivation	Impossible to grow.

Family Schizaeaceae

Description

Unusual slender fern with a short, slowly creeping, subterranean rhizome and tufts of extremely narrow, wiry fronds. The sterile fronds, to 10 cm long, are unbranched, whereas the taller fertile fronds are often forked at least once. At their apex the fertile fronds bear comb-like structures which carry the spores. Initially these are green, but they quickly age to brown after the spores are shed.

Distribution	Eastern bayside suburbs, and disjunctly distributed in southern areas of the state; also Qld, NSW, Tas, NZ and NCal.
Habitat	Coastal scrub, heathland, heathy forest and open forest.
Notes	This fern produces flushes of new fronds in spring and early summer and is dormant over the rest of the year. *S. asperula* is a synonym.
Similar species	*S. fistulosa* has undivided fronds.
Cultivation	Impossible to grow.

223

Family Selaginellaceae

Description

Small fern-ally spreading by straw-coloured subterranean stolons and with tufts of slender, erect, branched stems, each resembling a miniature conifer, the branches all arising in one plane. These are clothed with tiny, scale-leaves which are usually yellowish-green, but turn red when the plants dry out.

Distribution	Eastern and south-eastern suburbs, and widespread in southern areas of the state; also Qld, NSW, Tas and NT.
Habitat	Moist to wet depressions in peaty soil in heathland, heathy forest and open forest.
Notes	This species sometimes grows in large patches.
Similar species	*S. gracillima* has tiny stems which are unbranched, or only branch once or twice.
Cultivation	Impossible to grow.

Family Asteraceae

Description

Perennial herb forming a rounded clump, with soft, pithy stems, to 50 cm tall, and dark green leaves, to 7 cm long, which are usually toothed or pinnately lobed. The stems end in yellow, daisy flower-heads, 1.5–2 cm across, with numerous radiating florets.

Flowering period	Sept. to March.
Distribution	Northern and eastern suburbs, and throughout the state; occurs in all states and NZ.
Habitat	Open forest, woodland and coastal scrubs.
Notes	An extremely variable species, especially in growth habit, leaf shape and leaf division, with plants from some areas having small leaves with entire margins.
Similar species	*S. spathulatus* has an open habit with weak, spreading stems.
Cultivation	Easily grown in a range of soil types and positions.

Family Asteraceae

Description

Perennial herb forming bushy clumps, to 1.5 m tall, with a central rootstock and erect stems densely clothed with fleshy, dark green leaves, to 15 cm x 4 cm, with entire to coarsely toothed margins. Yellow, daisy flower-heads, about 1 cm across, are borne in conspicuous, flat-topped clusters at the end of each stem. Each flower-head usually has five spreading florets.

Flowering period	Nov. to Feb.
Distribution	Eastern suburbs, Dandenong Ranges, and widespread in cooler parts of the state; also NSW, ACT and Tas.
Habitat	Moist slopes and valleys in open forest and woodland.
Notes	A robust species which often colonises disturbed areas and is also prominent after fires.
Similar species	None in the region.
Cultivation	Readily grown in a moist sheltered position.

Family Solanaceae

Description

Fast-growing bushy shrub with fleshy angular stems clothed with soft, dark green leaves, to 30 cm x 2 cm. The lower leaves are commonly larger and deeply lobed, whereas the upper leaves are often entire. Purple or violet-coloured flowers, 2.5–4 cm across, are carried in axillary clusters, and followed by large, fleshy, egg-shaped fruit which ripen orange to red.

Flowering period	Sept. to March.
Distribution	Northern and eastern suburbs, and in eastern parts of the state; also Qld, NSW, Tas and NZ.
Habitat	Open forest, woodland and heathy forest.
Notes	This species often colonises disturbed areas and is also prominent after fire. The green fruit is reputedly poisonous.
Similar species	*S. laciniatum* has rounder, shallowly lobed flowers and yellow to orange fruit.
Cultivation	A fast-growing species which adapts readily to most sites. Plants respond well to heavy pruning.

Family Epacridaceae

Description

Slender shrub with stiffly erect stems, to 1.5 m tall, clothed with rigid, sharply pointed, green or yellowish leaves, to 1.2 cm x 0.3 cm, which are concave at the base. The stems end in dense, broad clusters of pink, starry flowers.

Flowering period	Aug. to Dec.
Distribution	Mainly south-eastern suburbs, and widespread in southern parts of the state; also NSW, Tas and SA.
Habitat	Moist to wet depressions and swamps in heathland and heathy forest.
Notes	In suitable conditions this species forms dense stands which are colourful when in flower.
Similar species	None in the region.
Cultivation	An excellent plant for containers, but can be difficult to establish in the garden.

Family Rhamnaceae

Description

Rounded or spreading shrub, to 3 m tall, with a dense habit and strongly-veined, blunt or notched, oval leaves, to 2.5 cm x 0.8 cm, which are dark green on the upper surface and paler and hairy beneath. Flat clusters of small, white flowers terminate the branchlets, with each cluster being surrounded by whitish leaves, which look as though they have been dusted with flour.

Flowering period	July to Nov.
Distribution	Northern and eastern suburbs, Dandenong Ranges, and throughout the state; also NSW, Tas and SA.
Habitat	Moist slopes and valleys in open forest.
Notes	This species may colonise disturbed areas and is often prominent on embankments and road verges.
Similar species	None in the region.
Cultivation	Readily grown in well-drained but moist soils in a sheltered position.

Family Stackhousiaceae

Description

Perennial herb with a thickened rootstock and erect, reddish stems, to about 30 cm tall, sparsely clothed with narrow, dark green leaves, to 4 cm long. Small, tubular, cream to white flowers are carried in long cylindrical spikes at the end of each stem.

Flowering period	Aug. to Jan.
Distribution	Widespread in the Melbourne region and throughout much of the state; also Qld, NSW, ACT, Tas, SA.
Habitat	Moist grassy areas in open forest, woodland and grassland.
Notes	This species, which often grows in colonies, is very attractive when in flower. Each plant can flower over a long period and the flowers are perfumed towards dusk and at night.
Similar species	*S. spathulata,* which grows in coastal dunes, has thicker broader leaves.
Cultivation	A useful plant in moist but well-drained soil.

Family Caryophyllaceae

Description

Perennial herb spreading by suckers and forming clumps of hairy, tangled stems clothed with whitish old leaves and clusters of narrow, bright green, prickly leaves, to 1.5 cm long. Starry white flowers, about 2 cm across, are borne singly on slender stalks from the upper axils.

Flowering period Mainly Oct. to Dec., but also sporadic at other times.

Distribution Northern and eastern suburbs, Dandenong Ranges, and throughout the state; also Qld, NSW, ACT, Tas and SA.

Notes This species may form untidy tangled mats littered with patches of dead foliage. Clumps regenerate strongly after fire.

Similar species None in the region.

Cultivation Grows readily in moist well-drained soil. Withstands heavy pruning.

Family Stylidiaceae

Description

Perennial herb with a dense basal tussock of erect, narrow, greyish-green to dark green leaves, to 30 cm x 1 cm, and upright, slender flower-spikes, to 50 cm tall, which carry numerous, small, bright pink flowers. Each flower has two opposite pairs of rounded petals, and an unusual central column-like structure which carries the pollen and stigma.

Flowering period	Sept. to Dec.
Distribution	Northern and eastern suburbs, Dandenong Ranges, and widespread throughout the state; also Qld, NSW, ACT, Tas and SA.
Habitat	Grassy areas in open forest, woodland and grassland.
Notes	A familiar plant which often grows in localised patches. The column in the centre of the flower is triggered by touch, leaving pollen on the back of any visiting insect. Children delight in triggering this mechanism.
Similar species	None in the region.
Cultivation	Readily grown in moist but well-drained soil.

232

Family Aizoaceae Climbing Spinach, Bower Spinach

Description

Robust climber with long slender stems which spread across the ground forming a mat or thread vigorously through surrounding shrubs to form a curtain of growth. Each stem is liberally covered with bright green, thick, fleshy diamond-shaped leaves, to 8 cm x 5 cm. Small yellow flowers, about 0.8 cm across, borne on long stalks among the leaves, are followed by small fleshy red to blackish berries.

Flowering period	Sporadic all year.
Distribution	Eastern and western bayside suburbs and in coastal areas of the state; also Tas, SA and WA.
Habitat	Coastal scrubs and saltmarshes.
Notes	An important species for stabilising coastal dunes and headlands. The leaves and young stems can be cooked as a substitute for spinach and were eaten by the Aborigines.
Similar species	*T. tetragonioides* is a smaller prostrate shrub with green flowers and ribbed fruit.
Cultivation	Easily grown in a sunny position in well-drained soil.

233

Family Tremandraceae

Description

Small shrub, to 50 cm tall, forming erect to spreading clumps, with thin, wiry stems and small, oval, hairy, dark green leaves, to 1.2 cm long, arranged in whorls of three or four. Pendulous, bell-like flowers, usually pink with a blackish centre, are produced in profusion from the upper axils.

Flowering period	July to Dec.
Distribution	Eastern and south-eastern suburbs, Dandenong Ranges, and widespread in the state; also NSW, Tas and SA.
Habitat	Open forest and woodland.
Notes	This species produces profuse displays of colourful flowers.
Similar species	*T. bauerifolia* has whorls of four to six narrower leaves with the margins curved under.
Cultivation	Requires well-drained soil in filtered sun or semi-shade. An attractive container plant.

Family Phormiaceae

Description

Perennial, lily-like herb which develops a large, erect tussock of coarse green to bluish-green leaves, to 30 cm long, each folded along the midrib to appear v-shaped in section. Wiry, branched flower-stems, much longer than the leaves, carry blue starry flowers, about 2.5 cm across. Each flower has six spreading segments and prominent yellow anthers. The fruit are small, greenish capsules.

Flowering period	Sept. to March.
Distribution	Eastern and south-eastern suburbs, Dandenong Ranges, and throughout lowland areas of the state; also Qld, NSW, Tas and SA.
Habitat	Heathland, heathy forest, open forest and woodland, often in moist soils.
Notes	Occasional plants have cream or white flowers. Each flower lasts a single day.
Similar species	Species of *Dianella* have similar flowers which are followed by bright blue, fleshy fruit.
Cultivation	Readily grown in a container or in well-drained soil in the garden.

Family Orchidaceae

Description

Terrestrial orchid with a slender cylindrical leaf, to 12 cm x 0.3 cm, dark green with a reddish base, and a slender, zig-zagged flower-stem, to 25 cm tall, bearing one to four flowers, 2.5–3 cm across. These flowers, which open widely in the sun, are yellow with two prominent, dark brown, ear-like structures on the central column.

Flowering period	Sept. and Oct.
Distribution	Eastern and south-eastern suburbs, and mainly in western parts of the state; also Tas, SA and WA.
Habitat	Open forest, heathy forest and heathland.
Notes	A distinctive orchid which grows in sparse colonies. The flowers, which open only in sunny weather, are fragrant.
Similar species	None in the region.
Cultivation	There has been little success in cultivating this orchid.

Family Orchidaceae

Description

Terrestrial orchid with a narrow leaf, to 15 cm x 0.3 cm, dark green with a reddish base, and a thin, zig-zagged flower stem, to 25 cm tall, bearing one to four flowers, about 1.5 cm across. These flowers, which only open widely in hot weather, are pink to reddish, with a prominent yellow patch on the central column.

Flowering period	Oct. and Nov.
Distribution	Eastern and south-eastern suburbs, Dandenong Ranges, and scattered in the state; also Qld, NSW, ACT, Tas and SA.
Habitat	Open forest, heathy forest and heathland.
Notes	In long periods of cool weather the flowers of this species may remain closed and can self-pollinate without opening.
Similar species	*T. rubra* has paler pink flowers about 2.5 cm across.
Cultivation	Grown in containers by orchid specialists.

Family Orchidaceae

Description

Terrestrial orchid with a large, thick, fleshy, dark green, ribbed leaf, to 30 cm x 2 cm, and a stout flower-stem, to 90 cm tall, which bears five to 30 flowers, 2.5–3 cm across, in a dense showy raceme. These flowers, which only open widely in sunny weather, are bright blue, and the central column has a blackish colour band below the yellowish top, and two white hair tufts.

Flowering period	Oct. to Dec.
Distribution	Eastern suburbs, Dandenong Ranges, and in eastern parts of the state; also NSW.
Habitat	Moist areas in open forest and tall forest.
Notes	A handsome orchid with impressive colourful displays of flowers.
Similar species	*T. aristata,* from coastal districts, has a longer central column without a colour band.
Cultivation	There has been little success in cultivating this orchid.

Family Poaceae

Description

Perennial grass forming a dense tussock of erect to limp, green, brownish or bluish leaves, to 30 cm x 0.3 cm. Old tussocks often contain a high proportion of dead leaves. Slender wiry flower-stems, to 80 cm tall, carry dense clusters of shiny brown spikelets, with the fertile spikelet in each group bearing a long awn.

Flowering period	Mainly Sept. to Feb., but also sporadic after rain.
Distribution	Widespread in the Melbourne region and throughout the state; occurs in all states and many overseas countries.
Habitat	Grassland, open forest and woodland.
Notes	A distinctive grass which often forms extensive pure swards. Tussocks recover vigorously from fire. It was formerly well known as *T. australis.*
Similar species	None in the region.
Cultivation	Readily grown in a wide range of soils and positions.

Family Anthericaceae

Description

Perennial, bulbous, lily-like herb with a sparse tussock of erect, very narrow, greyish leaves, to 20 cm long, and upright wiry flower-stems, to 30 cm tall, each topped with an open-branched cluster of buds and flowers. Each flamboyant flower, which is 2.5–3.5 cm wide, has three narrow sepals and three broad purple petals, which are delicately fringed with spreading, fine, purple hairs.

Flowering period	Mainly Nov. to Jan., but sometimes later.
Distribution	Widespread in the Melbourne region and throughout much of the state; also Qld, NSW, ACT and SA.
Habitat	Grassy forests, open forest and woodland.
Notes	This species has exquisite, brightly coloured flowers which unfortunately last less than a day.
Similar species	*T. patersonii* is a climber with much smaller flowers.
Cultivation	Successful in containers, but can be difficult to establish in the garden.

Family Anthericaceae

Description

Perennial, lily-like herb with fleshy roots and wiry stems which branch freely to form a tangled clump, to about 50 cm tall. The plants have a few narrow basal leaves, which are often withered by flowering time, and the stems support masses of starry yellow flowers, each 1–1.8 cm across. These flowers, which open in mid-morning and close by late afternoon, have six widely spreading segments and a prominent central cluster of stamens. The buds are brownish and withered flowers twist in a spiral manner.

Flowering period	Oct. to April, but also sporadic.
Distribution	Widespread in the Melbourne region and throughout much of lowland Vic; also Qld, NSW, Tas, SA, WA and NT.
Habitat	Grasslands, grassy forest and heathland.
Notes	This species produces impressive floral displays especially in wet years. It can withstand some disturbance.
Similar species	None in the region.
Cultivation	Readily grown in containers or a well-drained position in the garden.

Family Juncaginaceae

Description

Perennial aquatic herb with a thick fleshy rhizome, bearing tubers and a tussock of ribbon-like, bright green leaves, 1–2 m long and to about 3 cm wide. These float on water or form a semi-erect tussock in drying mud. Tiny greenish flowers are massed in dense cylindrical spikes, which are held erect above the water, and are followed by small, globular, green fruit.

Flowering period	Aug. to April.
Distribution	Widespread in the Melbourne region and throughout the state; occurs in all states.
Habitat	Shallow fresh water in lakes, dams and billabongs and also in slow moving streams.
Notes	This species usually grows in colonies. The root tubers are edible and were an important food source for the Aborigines.
Similar species	*T. striata* is much smaller with very narrow leaves.
Cultivation	Readily grown as an aquatic in dams or ponds.

Family Goodeniaceae

Description

Perennial herb with a persistent, slender rootstock and basal rosette of broad, light-green leaves, to 6 cm x 1.5 cm, with bluntly toothed margins. Slender, wiry flower-stems, to 50 cm tall, branch widely near the apex, each branch ending in a bright golden-yellow flower, about 2 cm across, with a prominent basal spur.

Flowering period	Oct. to Dec. and March to May.
Distribution	Western and northern suburbs, and widespread in the state; also Qld, NSW, ACT, Tas and SA.
Habitat	Moist grassy areas in open forest, woodland and grassland.
Notes	This species often grows in localised colonies but is occasionally seen in extensive patches.
Cultivation	Readily grown in moist soil in a sunny position.

Family Menyanthaceae

Description

Perennial herb forming clumps, or if growing in water then spreading by vigorous stolons and with floating leaves. Clumping plants have an erect rosette of shiny, dark green, kidney-shaped leaves, to 10 cm x 6 cm, borne on long stems. Erect flower-stems, to 1 m tall, carry clusters of bright yellow flowers, each 1–1.5 cm across.

Flowering period	Mainly Sept. to Dec., but also sporadic at other times.
Distribution	Eastern suburbs, and widespread throughout the state; also NSW, Tas and SA.
Habitat	Swamps and wet depressions in open forest, woodland and grassland.
Notes	This species spreads rapidly after flooding to colonise suitable wet sites.
Similar species	*V. exaltata* has oval leaves.
Cultivation	Readily grown in a bog garden or shallow pond.

Family Violaceae

Description

Perennial herb forming sparse to dense clumps of erect dark green leaves to about 25 cm long. Each leaf has a long narrow stalk and a broad blade, to about 8 cm x 6 cm. Showy blue to purplish violet-shaped flowers, 1.5–2 cm across, are borne singly on slender stalks about as long as the leaves.

Flowering period	Sept. to Feb.
Distribution	Northern and eastern suburbs, Dandenong Ranges and in many parts of the state; also Qld, NSW, ACT, Tas and SA.
Habitat	Grassy areas of open forest, woodland and heathy forest.
Notes	At peak flowering plants produce an attractive floral display. A form from Ringwood has a compact growth habit and smaller than normal leaves.
Similar species	None in the region.
Cultivation	Easily grown in a sheltered position in moisture retentive soil. An attractive rockery plant which may naturalise from seed.

Family Violaceae

Description

Perennial herb spreading by vigorous stolons and forming a dense mat, with masses of light green, kidney-shaped leaves on the end of slender stalks. White, violet-shaped flowers, about 1 cm across, with purple centres, are carried high above the leaves on thin stalks.

Flowering period	Mainly June to March, but also sporadic at other times.
Distribution	Northern and eastern suburbs, Dandenong Ranges, and throughout much of the state; also Qld, NSW, ACT, Tas and SA.
Habitat	Moist areas in open forest and woodland.
Notes	A robust species which quickly colonises suitable sites.
Similar species	*V. sieberiana* has much smaller leaves with coarsely toothed margins.
Cultivation	Readily grown in a moist sheltered position.

Family Campanulaceae

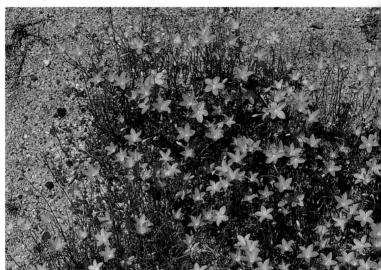

Description

Perennial herb spreading by stolons and forming clumps of erect, thin, wiry stems, to 50 cm tall, which are sparsely covered with narrow, dark green, hairy leaves, to 4 cm x 0.3 cm, which can have toothed margins. Bluebell-shaped flowers, to 1.2 cm across, are well-displayed at the end of slender stems.

Flowering period	Mainly Nov. to May, but also sporadic at other times.
Distribution	Widespread in the Melbourne region and throughout the state; also Qld, NSW, ACT, SA, WA and NT.
Habitat	Grassy areas in open forest, woodland and grassland.
Notes	A showy species which may colonise disturbed areas and is frequently seen growing on road verges.
Similar species	*W. stricta* has the lower leaves with wavy margins and pale blue to whitish flowers.
Cultivation	Readily grown in moist but well-drained soils.

Family Colchicaceae

Description

Perennial, bulbous, lily-like herb with three very narrow, bright green, grass-like leaves, to 10 cm long, which expand greatly at the base to sheath the flower-stem. Male and female flowers are borne on separate plants, or occasionally bisexual flowers can occur. The flowers, each about 1 cm across, are white with a prominent transverse purple band towards the base of each segment. Male flowers have prominent red or purple anthers, and the females have a swollen central ovary.

Flowering period	Aug. to Oct.
Distribution	Widespread in the Melbourne region and over much of the state; also Qld, NSW, ACT, Tas and SA.
Habitat	Moist grassland, grassy forests, open forest, woodland and heathland.
Notes	The flowers release a strong perfume which is especially noticeable towards dusk.
Similar species	None in the region.
Cultivation	Difficult to establish.

Family Xanthorrhoeaceae

Description

Perennial plant with a thick woody, often branched trunk, to about 3 m tall, each branch being topped with a rounded crown of stiff, thin, 4-sided, arching, bluish-green leaves. On unburnt plants, a prominent skirt of hanging, whitish, dead leaves is also present. Masses of creamy-white, scented flowers are borne in dense, cylindrical spikes, these can sometimes be more than 2 m long.

Flowering period	July to Dec.
Distribution	North-eastern and south-eastern suburbs, *but now rare in the region*; widespread in the state, also NSW, Tas and SA.
Habitat	Open forest, heathy forest and heathland.
Notes	This species produces spectacular flowering displays in the first season after fires. The flowers are rich in nectar and attract birds and a wide range of insects.
Similar species	*X. minor* has a subterranean trunk and three-sided leaves.
Cultivation	Very slow growing. Requires excellent drainage in a sunny position.

Family Xyridaceae

Description

Perennial herb forming large, dense clumps, with stiff, narrow, cylindrical leaves, to 60 cm long, and wiry flower-stems, to 1.5 m tall. Each ends in a dense cluster of dark brown bracts from which bright yellow flowers, about 2 cm across, are produced at intervals. Each flower, which has three broad, flimsy petals, lasts less than a day.

Flowering period	Nov. to Feb.
Distribution	Once widespread in eastern bayside suburbs, *but now very rare or extinct*; widespread in southern parts of the state, also Qld, NSW, Tas and SA.
Habitat	Swampy areas in heathland and heathy forest.
Notes	The large colourful flowers are produced at intervals throughout the flowering season.
Similar species	*X. gracilis* grows to about 1 m tall and the floral bracts have a pale central area.
Cultivation	An interesting species for growing in a container.

Family Rutaceae

Description

Vigorous shrub, to 4 m tall, with opposite pairs of long-stalked, trifoliolate leaves which have dark green leaflets, to 10 cm x 1.5 cm. White, four-petalled flowers, about 1 cm across, borne in large, flat-topped clusters from the upper axils, produce a prominent floral display.

Flowering period	Aug. to Dec.
Distribution	Dandenong Ranges, and southern areas of the state; also NSW and Tas.
Habitat	Cool moist slopes and gullies in tall forest.
Notes	All parts of this species have a strong smell which is obnoxious to some people, even to the extent of inducing headaches.
Similar species	None in the region.
Cultivation	Readily grown in a sheltered position.

Family Zygophyllaceae

Description

Annual herb which forms sprawling or scrambling clumps of fleshy, ridged stems with paired, succulent, pale green to bluish-green leaflets attached to a single stalk in a Y-shape. Bright yellow flowers, 1–1.4 cm across, are borne on short stalks in the upper leaf axils.

Flowering period	Aug. to Dec.
Distribution	Northern suburbs, *where sporadic to rare*, and in north-western parts of the state; also Qld, NSW, SA, WA and NT.
Habitat	Among rocks and open sites near streams.
Notes	An annual species which grows rapidly and dies after seeds reach maturity.
Similar species	*Z. billardieri* is a robust perennial with dark green leaves.
Cultivation	Requires a sunny position in well-drained soil.

Glossary

alternate	Arranged at different levels on a stem.
annual	A plant which grows, flowers, sets seed and dies in one year.
anthers	The part of a stamen which bears pollen.
apex (apical, adj.)	The tip or point of a leaf.
awn	A bristle-like projection, usually found on the flowers or seeds of grasses.
axil	The angle formed between a leaf and its supporting stem.
axillary	Arising in the axil.
basal	At or from the base.
bipinnate	Leaves twice divided.
bisexual	Of two sexes.
blade	The expanded part of a leaf.
box	Group of eucalypts with persistent, short-fibred, tessellated bark.
bract	Leaf-like structure which occurs on a flower-stem.
bulbil	Tiny plant produced at the junction of main veins on the fronds of some ferns.
bulbous	Swollen or bulb-shaped.
burr	A prickly, usually globular, fruit.
callus (i)	Specialised structure(s) developed on the lip of orchids.
calyx	All of the sepals.
capsule	A dry fruit which splits or opens to release seeds.
clone	A group of vegetatively propagated plants with a common ancestry.
column	A fleshy structure containing the sex organs, found in the centre of orchid flowers.
cone	A woody fruit found in conifers.
congested	Crowded, close together.
dimorphic	In two distinct forms.
dissected	Deeply divided into segments.
dune	A mound formed from wind-blown sand.
elliptical	An elongated oval with matching rounded sides.
elongate	Drawn out in length.
endemic	Restricted to a particular country, region or area.
entire	Not toothed, lobed or dissected.
epiphyte	A plant growing on another plant, but not a parasite.
exserted	Protruding, jutting out.
family	A grouping of related genera.
filament	The stalk of a stamen supporting the anther.
floret	A single flower in a compound head of flowers.
flower-head	A compound head of stalkless flowers.
forb	Herbs other than grasses, sedges and rushes.
frond	Leaf of a fern.
fruit	A seed-bearing organ.
genus (pl. genera)	A grouping of related species
glabrous	Without hairs.
glaucous	Bluish-green, often as the result of a powdery coating.
globose	Globular, almost spherical.
grassland	Treeless areas dominated by grasses and forbs.
habit	The general appearance of a plant.
heathland	Dense shrubby vegetation dominated by tough, wiry-stemmed, small-leaved plants.
heathy forest	Forest dominated by trees with usually short, crooked trunks and the groundcover is densely shrubby, with many heathy species.
herb	A plant with fleshy, not woody, stems; sometimes strictly applied to grasses, rushes and sedges (see also forb).
indigenous	Native to a country, region or area.
inflorescence	The flowering arrangement of a plant.
juvenile	The young stage of growth before a plant is capable of flowering.
leaflet	A segment of a compound or divided leaf.
lignotuber	Woody swelling containing dormant buds at the base of a stem or trunk.

lip	The central petal of an orchid which is often highly modified.
lobed	Divided into segments.
margin	The edge.
membranous	Thin-textured.
midrib	The main vein that runs the full length of a leaf.
node	Point on the stem where leaves, bracts or flowers arise.
once-divided	With one level of division.
open forest	Forest in which the canopy is almost continuous and is dominated by trees with well-developed trunks and underneath is a grassy or shrubby groundcover.
opposite	Arising at opposite sides but at the same level.
ovate	Egg-shaped in flat section.
panicle	A branched raceme.
parasite	A plant which draws nourishment from another.
pendent	Hanging down.
perennial	Living for several years.
petal	One segment of the corolla or inner whorl of the flower.
petiole	The stalk of a leaf.
phyllode	Modified leaf stalk acting as a leaf.
pinnate	Once-divided, with the divisions extending to the midrib.
plantlet	A tiny plant.
pod	A dry fruit that splits when ripe to release its seeds.
prostrate	Lying flat on the ground.
raceme	Unbranched inflorescence with stalked flowers.
recurved	Curved backwards.
rhizome	An underground stem.
rootstock	A swollen taproot.
rosette	A group of leaves which radiate from near the base of a stem.
scale	A small dry rudimentary leaf.
scaly	Bearing scales.
sepal	One segment of the calyx or outer whorl of the flower.
sessile	Without a stalk.
simple	Undivided.
spike	Unbranched inflorescence with stalkless flowers.
stamen	Male part of a flower, consisting of an anther and a filament.
stolon	Subterranean stem or shoot rooting at intervals.
style	Elongated part of the carpel between the ovary and the stigma.
subspecies	A subgroup within a species to differentiate geographically isolated variants.
succulent	Fleshy or juicy.
sucker	A shoot arising from the roots.
synonym	A scientific name previously used for a species and not currently accepted.
tall forest	Moist to wet forest dominated by tall trees with many shrub layers.
taproot	The perpendicular main root of a plant.
tea-tree heath	Specialised coastal heath dominated by coastal tea-tree (*Leptospermum laevigatum*).
terminal	At the apex or end.
terrestrial	Growing in the ground.
tessellated	In a chequered or mosaic pattern.
trifoliolate	A divided leaf with three leaflets.
tuber	The swollen end of an underground root or stem.
tuberous	Swollen and fleshy, or bearing a tuber(s).
twiner	Climbing plant with spirally twining stems.
unisexual	Of one sex only.
valve	A segment on a woody fruit.
variety	A subgroup within a species to differentiate minor variants.
vein	The conducting tissue of leaves.
venation	The pattern formed by veins.
whorl	Three or more segments arising at one point.
woodland	Forest dominated by trees but the trees are spaced and do not form a continuous canopy. The ground layer may be grassy or shrubby.
woolly	With long, soft, matted hairs.

References and further reading

Backhouse, G.N. and Jeanes, J. (1995). *The Orchids of Victoria,*
Melbourne University Press.

Brooker, M.I H. and Kleinig, D.A. (1983). *Field Guide to Eucalypts* Vol. 1,
Inkata Press, Melbourne and Sydney.

Cochrane, R.G., Fuhrer, B.A., Rotherham, E.R. and Willis, J.H. (1968).
Flowers and Plants of Victoria, A. H. & A. W. Reed, Sydney.

Costermans, L. (1981). *Native Trees and Shrubs of South-Eastern Australia,*
Rigby, Adelaide.

Duncan, B.D. & Isaac, G. (1986). *Ferns and Allied Plants of Victoria,
Tasmania and South Australia,* Melbourne University Press, Carlton,
& Monash University, Clayton.

Eddy, D., Mallinson, D., Rehwinkel, R. and Sharp, S. (1999).
Grassland Flora, NCP, Canberra.

Flora of Australia Vol. 48, Ferns, Gymnosperms and Allied Groups.
Melbourne: ABRS/CSIRO, Australia (1998).

Foreman, D.B. and Walsh, N.G. (1993). *Flora of Victoria* Vol. 1,
Inkata Press, Melbourne.

Harden, G. (1990–93). *Flora of New South Wales* Vols 1-4,
University of New South Wales Press, Kensington.

Marriott, N. & J. (1998). *Grassland Plants of South-Eastern Australia,*
Bloomings Books, Melbourne.

Society for Growing Australian Plants (1991). *Flora of Melbourne,*
SGAP Maroondah, Inc.

Society for Growing Australian Plants (1995). *Plants of Melbourne's Western
Plains,* SGAP Keilor Plains Group.

Walsh, N.G. and Entwisle, T.J. (1994). *Flora of Victoria* Vol. 2,
Inkata Press, Melbourne.

Willis, J.H. (1962–72). *A Handbook to Plants in Victoria* Vols 1 & 2,
Melbourne University Press, Melbourne.

INDEX